T0260541

Annals of Mathematics Studies

Number 33

ANNALS OF MATHEMATICS STUDIES

Edited by Emil Artin and Marston Morse

1. Algebraic Theory of Numbers, *by* HERMANN WEYL
3. Consistency of the Continuum Hypothesis, *by* KURT GÖDEL
6. The Calculi of Lambda-Conversion, *by* ALONZO CHURCH
7. Finite Dimensional Vector Spaces, *by* PAUL R. HALMOS
10. Topics in Topology, *by* SOLOMON LEFSCHETZ
11. Introduction to Nonlinear Mechanics, *by* N. KRYLOFF *and* N. BOGOLIUBOFF
14. Lectures on Differential Equations, *by* SOLOMON LEFSCHETZ
15. Topological Methods in the Theory of Functions of a Complex Variable, *by* MARSTON MORSE
16. Transcendental Numbers, *by* CARL LUDWIG SIEGEL
17. Problème Général de la Stabilité du Mouvement, *by* M. A. LIAPOUNOFF
19. Fourier Transforms, *by* S. BOCHNER *and* K. CHANDRASEKHARAN
20. Contributions to the Theory of Nonlinear Oscillations, Vol. I, *edited by* S. LEFSCHETZ
21. Functional Operators, Vol. I, *by* JOHN VON NEUMANN
22. Functional Operators, Vol. II, *by* JOHN VON NEUMANN
23. Existence Theorems in Partial Differential Equations, *by* DOROTHY L. BERNSTEIN
24. Contributions to the Theory of Games, Vol. I, *edited by* H. W. KUHN *and* A. W. TUCKER
25. Contributions to Fourier Analysis, *by* A. ZYGMUND, W. TRANSUE, M. MORSE, A. P. CALDERON, *and* S. BOCHNER
26. A Theory of Cross-Spaces, *by* ROBERT SCHATTEN
27. Isoperimetric Inequalities in Mathematical Physics, *by* G. POLYA *and* G. SZEGO
28. Contributions to the Theory of Games, Vol. II, *edited by* H. KUHN *and* A. W. TUCKER
29. Contributions to the Theory of Nonlinear Oscillations, Vol. II, *edited by* S. LEFSCHETZ
30. Contributions to the Theory of Riemann Surfaces, *edited by* L. AHLFORS *et al.*
31. Order-Preserving Maps and Integration Processes, *by* EDWARD J. MCSHANE
32. Curvature and Betti Numbers, *by* K. YANO *and* S. BOCHNER
33. Contributions to the Theory of Partial Differential Equations, *edited by* L. BERS, S. BOCHNER, *and* F. JOHN

CONTRIBUTIONS TO THE THEORY OF PARTIAL DIFFERENTIAL EQUATIONS

S. BERGMAN
L. BERS
S. BOCHNER
F. E. BROWDER
J. B. DIAZ
A. DOUGLIS
F. JOHN
P. D. LAX
J. LERAY
C. LOEWNER
A. N. MILGRAM
C. B. MORREY
L. NIRENBERG
M. H. PROTTER
P. C. ROSENBLOOM

Edited by
L. Bers, S. Bochner, and F. John

The papers in this volume were read at the Conference on Partial Differential Equations sponsored by the National Academy of Sciences—National Research Council, October 1952.

Princeton, New Jersey
Princeton University Press
1954

London: Geoffrey Cumberlege, Oxford University Press
L. C. Card 54-5004

Printed in the United States of America

FOREWORD

In October 1952 a three day conference on partial differential equations was held at Arden House, Harriman, New York. The conference was organized and sponsored by the National Academy of Sciences - National Research Council.

This volume contains those papers, read at the conference, which were submitted by the authors for publication. The editors regret the unavoidable delay in publication and hope that this volume will prove to be useful to mathematicians working in this field.

The editing and preparing of this study was carried out entirely by Anneli Lax. The editors gratefully acknowledge her valuable assistance.

L. Bers
S. Bochner
F. John

CONTENTS

I. GREEN'S FORMULA AND ANALYTIC CONTINUATION

S. Bochner

For anyalytic functions in more than one complex variable there is a theorem of Hartog's that if a function is given on the connected boundary of a bounded domain, then it can be continued analytically into all of the domain. The class of functions to which this theorem applies was considerably generalized in our paper [1]: "Analytic and meromorphic continuation by means of Green's formula," Annals of Mathematics 44 (1943), 652-673; and it was further expanded in our recent note [2]: "Partial differential equations and analytic continuation," Proceedings of the National Academy of Sciences 38 (1952), 227-30. Now, in §1 of the present paper the leading theorem of [2] will be given its final version known to us (see Theorem 5) and, furthermore, details of the proof will be modified and added.

The real and imaginary parts of analytic functions of complex variables are solutions of a system of Cauchy-Riemann equations in real variables. In the case of more than one complex variable, this system is quite complicated and, as it turns out, much too restrictive for our theorem. At first in [1], and then more systematically and generally in [2], we introduced instead a system consisting of only two equations, both with constant coefficients: an elliptic one in all variables and some other one in fewer than all variables; the second equation was the one by which the actual continuation was brought about. However, the second equation could only operate if the function was first represented by a certain Green's formula, and it was the sole task of the elliptic equation to secure just such a formula. Now, in further analyzing certain aspects of our theorem, we found it pertinent to try to give up the elliptic equation altogether and to hypothesize directly a Green's formula having the requisite properties. This will be done in the present paper.

In §2 we will be dealing in a similar fashion with another theorem in several complex variables which although closely related to the previous one is different from it nevertheless. Following up a suggestion of Severi's, this theorem was presented more systematically than had been done before in Chapter IV of the book by Bochner-Martin: Several Complex Variables, Princeton, 1948. It will now be given a rather more general version than previously.

§1. EUCLIDEAN SPACES

In Euclidean E_n: $(\xi^1, \ldots, \xi^n)$ we take a p-form, $1 \le p \le n-1$

(1) $$\frac{1}{p!} \sum_{(\alpha)} A_{\alpha_1 \ldots \alpha_p} d\xi^{\alpha_1} \ldots d\xi^{\alpha_p}$$

and we assume that each component of the skew tensor is a finite linear partial differential expression involving an unspecified function $f(\xi)$ with coefficients which are functions of the difference $\xi - x = (\xi^\beta - x^\beta)$, where $x = (x^\beta)$ is another variable point of the given space. Thus we have

(2) $$A_{\alpha_1 \ldots \alpha_p} = \sum_{(\nu)} G^{\nu_1 \ldots \nu_n}_{\alpha_1 \ldots \alpha_p} (\xi - x) \wedge_{\nu_1 \ldots \nu_n} f(\xi)$$

where

(3) $$\wedge_{\nu_1 \ldots \nu_n} f(\xi) = \frac{\partial^{\nu_1 + \ldots + \nu_n} f(\xi)}{(\partial \xi^1)^{\nu_1} \ldots (\partial \xi^n)^{\nu_n}}$$

with $\nu_1 \ge 0, \ldots, \nu_n \ge 0$, $\nu_1 + \ldots + \nu_n \le N - 1$ for some N sufficiently large but finite. In this sense we denote the form (1) by

(4) $$G_p(\xi - x;\ f(\xi);\ d\xi)$$

and we stipulate that the individual functions $G^{(\nu)}_{(\alpha)}(t)$, $t = \xi - x$, which occur in (2) shall be defined and real analytic in a certain open set T of the Euclidean E_n: (t^β).

We now introduce the requirement

(5) $$d_\xi G_p \equiv 0$$

that is

(6) $$\sum_{q=1}^{p+1} (-1)^q \frac{\partial}{\partial \xi^{\alpha_q}} A_{\alpha_1 \ldots \alpha_{q-1} \alpha_{q+1} \ldots \alpha_{p+1}} = 0$$

and this is a system of equations

(7) $$\sum_{(\nu)} H^{\nu_1 \ldots \nu_n}_{\alpha_1 \ldots \alpha_{p+1}} (\xi - x) \wedge_{\nu_1 \ldots \nu_n} f(\xi) = 0$$

$$0 \le \nu_1 + \ldots + \nu_n \le N$$

in which the coefficients $H^{(\nu)}_{(\alpha)}$ are linear combinations of the coefficients $G^{(\nu)}_{(\alpha)}$ and their first derivatives.

We now introduce fictitiously an "elliptic operator" $\Delta_\xi f(\xi)$ and we will say that a function $f(\xi)$ which is defined and analytic in a domain U of E_n satisfies there the equation

(8) $$\Delta_\xi f(\xi) = 0$$

if the relations (7) are fulfilled for ξ in U and $\xi - x$ in T. For every U the space of solutions of (8) is a vector space having the following closure property: Let $f_1(\xi)$, $f_2(\xi)$, ... be a sequence of solutions. If every point ξ^β of U has a complex neighborhood into which all the functions of the sequence can be continued analytically, and if these functions converge uniformly in these neighborhoods, then the limit function is again a solution.

Furthermore, if we put $\xi = t + x$ in (7), we obtain

(9) $$\sum_{(\nu)} H^{(\nu)}_{\alpha}(t) \wedge^t_{\nu_1 \cdots \nu_n} f(t+x) = 0$$

and therefore, locally, if $f(\xi)$ is a solution of (7) then so is also the translated function $f(\xi+h)$ for a sufficiently small constant displacement $h = (h^\beta)$. This property is crucial to our purpose, and we express it symbolically by saying that our (fictitious) operator Δ has "constant coefficients"; and if we combine all properties enumerated, we obtain the further property that if $f(\xi)$ is a solution, then so are also all first derivatives $\wedge_\beta f(\xi)$, and hence, also all mixed partial derivatives, and hence, also every finite linear combination with constant coefficients

(10) $$\wedge f = \sum_{(\mu)} a_{\mu_1 \cdots \mu_n} \wedge_{\nu_1 \cdots \nu_n} f(\xi)$$
$$0 \leq \mu_1 + \cdots + \mu_n \leq M$$

this operator being formed literally, and not just only fictitiously or symbolically.

We now take in U a p-dimensional chain B_p and we form the integral

(11) $$g(x) = \int_{B_p} G(\xi - x;\, f(\xi);\, d\xi)$$

for those points x for which it is definable, that is, for the open set X which is such that for x in X and ξ in B_p, $\xi - x$ is in T.

If we hold B_p fixed, then $g(x)$ is analytic in T; and if we vary $f(\xi)$ and then introduce the functional

$$g(x) = L(f(\xi);\, x) \tag{12}$$

then the latter is distributive, that is,

$$L(c_1 f_1 + c_2 f_2;\, x) = c_1 L(f_1;\, x) + c_2 L(f_2;\, x). \tag{13}$$

Next, if two chains B_p, B'_p are homologous relative to $U - (x)$, then by (5) we have

$$\int_{B_p} = \int_{B'_p} \tag{14}$$

and to this extent the integral (11) is "independent of the path" if $f(\xi)$ is a solution of (8). But the decisive property is yet to be stated and it is as follows.

THEOREM 1. If B_p is a cycle then the functional (12) is commutative with translations and partial differentiations. That is, (12) implies

$$g(x+h) = L(f(\xi + h);\, x) \tag{15}$$

for small h locally, and also

$$\frac{\partial g}{\partial x^\beta} = L\left(\frac{\partial f}{\partial \xi^\beta};\, x\right) \tag{16}$$

and more generally

$$\Lambda_x\, g(x) = L(\Lambda_\xi f(\xi);\, x) \tag{17}$$

for any operator (10).

PROOF. We have

$$g(x+h) = \int_{B_p} G(\xi - x - h;\, f(\xi);\, d\xi) \tag{18}$$

and if we replace ξ by $\xi + h$ (18) is equal to

(19)
$$\int_{B'_p} G(\xi - x;\ f(\xi + h);\ d\xi)$$

where $B'_p = B_p - (h)$ results from B_p by a translation. If x is given then for sufficiently small h, B'_p is homologous to B_p, and by (14) the integral (19) is

$$\int_{B_p} G(\xi - x;\ f(\xi + h);\ d\xi) = L(f(\xi + h);\ x)$$

as claimed in (15).

Due to (13) the result just obtained implies that the integral (11) carries a difference quotient

$$\frac{1}{h}[f(\xi^1, \ldots, \xi^\beta + h, \ldots, \xi^n) - f(\xi^1, \ldots, \xi^\beta, \ldots, \xi^n)]$$

into the corresponding difference quotient for $g(x^\alpha)$. If now we let $h \longrightarrow 0$, then we obtain (16), as can be easily proved, and then also (17), as claimed.

Theorem 1 will suffice for our present purposes. However, for a certain type of conclusion that was attempted in [1] and [2] a partial generalization of Theorem 1 is required for the case in which B_p is not a cycle, and we will state the generalization without proof.

THEOREM 2. If we are given a symbolic equation (8) and an operator (10), then there exists a (p-1)-form

$$Y_{p-1}(\xi - x;\ f(\xi);\ d\xi)$$

in which the partial derivatives of $f(\xi)$ occurring are of order $\leq N - 1 + M - 1$, and which has the following property: If B_p is a p-chain and B_{p-1} is its boundary and if $f(\xi)$ satisfies (8) in a domain containing $B_p + B_{p-1}$, then we have

$$L(\wedge_\xi f(\xi);\ x) - \wedge g(x) = \int_{B_{p-1}} Y_{p-1}(\xi - x;\ f(\xi);\ d\xi)$$

and thus we again have (17) provided that the function $f(\xi)$ and its derivatives of order $\leq N - 1 + M - 1$ are zero on the boundary B_{p-1}.

From now on we will consider only the dimension $p = n - 1$, and we will assume that the domain T on which the coefficients $G^{(\nu)}_{(\alpha)}(t)$ are defined is the entire E_n except for the origin, so that $L(f(\xi); x)$ is defined for x in $E_n - B_{n-1}$.

THEOREM 3. If B_{n-1} bounds a small subdomain D^o of U then we have

(20) $$\oint_{B_{n-1}} G(\xi - x; f(\xi); d\xi) = \sum_{(r)} c_{r_1 \ldots r_n} \wedge_{r_1 \ldots r_n} f(x)$$

$$r_1 \geq 0, \ \ldots, \ r_n \geq 0$$

where $c_{r_1 \ldots r_n}$ are constants independent of $f(\xi)$.

PROOF. It follows from (14) that for the proof of (20) we may assume that D^o is a coordinate sphere $|\xi| < 2\rho$, and that even the double sphere $|\xi| < 4\rho$ is still contained in U, and that the points x are restricted to the small sphere $|x| < \rho$. Now, on replacing ξ by $\xi + x$ we obtain

$$g(x) = \oint_{B_{n-1} - (x)} G(\xi; f(\xi + x); d\xi)$$

$$= \oint_{B_{n-1}} G(\xi; f(\xi + x); d\xi)$$

and for sufficiently small ρ we may now insert the Taylor expansion

$$f(\xi + x) = \sum_{(s)} \frac{(\xi^1)^{s_1} \ldots (\xi^n)^{s_n}}{s_1! \ \ldots \ s_n!} \wedge_{s_1 \ldots s_n} f(x)$$

and integrate term by term. Hence the conclusion.

The theorem is of no practical importance to us, but it puts into a proper perspective the assumption we are now going to make that the right side in (20) shall be identically $f(x)$, that is,

(21) $$\oint_{B_{n-1}} G(\xi - x; f(\xi); d\xi) = f(x)$$

and we express this assumption by saying that our integral (11) is a Green's formula. The first consequence of this assumption is:

THEOREM 4. If B is a simplex (which may be

part of any chain over which we integrate) then the function

(22) $$g(x) = \int_B G(\xi - x;\ f(\xi);\ d\xi)$$

can be continued across B from either side and the difference of the two values in the vicinity of B is $f(x)$,

(23) $$g_+(x) - g_-(x) = \pm f(x)$$

the algebraic sign being determined by the orientation of B.

PROOF. (Cf. [1], p. 656-57.) Denoting one side of B as positive, we deform B into a simplex B' on its negative side but keep the edges fixed. By (14) we have for x on the positive side

$$g_+(x) = \int_B = \int_{B'}$$

and thus $g_+(x)$ can be continued a certain distance into the negative side. Denoting the continuation still by $g_+(x)$, we now have for points x in the domain bounded by $B' - B$

$$g_+(x) - g_-(x) = \int_{B'} - \int_B = \int_{B'-B}$$

and by (21) this is $\pm f(x)$ as claimed.

THEOREM 5. If we are given a Green's formula and if the coefficients $G^{(\nu)}_{(\alpha)}(t)$ in (2) are of slow growth at infinity in the sense that for some $r_o > 0$ we have

(24) $$\Lambda_{r_1 \ldots r_n} G^{\nu_1 \ldots \nu_n}_{\alpha_1 \ldots \alpha_{n-1}}(t) = O(|t|^{-n-1}), \qquad |t| \longrightarrow \infty$$

for

(25) $$r_1 \geq 0, \ldots, r_n \geq 0, \quad r_1 + \ldots + r_n \geq r_o$$

then, if D is a bounded domain having a connected boundary B_{n-1} and if U is a neighborhood of B_{n-1} and if an analytic function $f(x)$ in U satisfies the associated "elliptic equation"

(26) $$\Delta_x f(x) = 0$$

in all its n variables and an additional equation

$$\wedge f \equiv \sum_{(q)} a_{q_1 \cdots q_m} \wedge_{q_1 \cdots q_m} f(x) = 0 \quad (m<n)$$

with constant coefficients not all zero in fewer than all variables, then the function $f(x)$ has an analytic continuation into all of $D + U$.

PROOF. By Theorem 4, if we restrict U sufficiently, we can put

$$f(x) = g_+(x) - g_-(x)$$

where $g_+(x)$ exists and is analytic in $D + U$, $g_-(x)$ is analytic in $(E^n - D) + U$, and the assertion of the theorem will follow if we show that $g_-(x)$ has an analytic continuation into all of E^n.

Let us denote $g_-(x)$ by $\phi(x)$. We know the following about this function: It is defined and analytic in the exterior of a sphere

(27) $$|x| > R$$

and there exists an $r' > 0$ such that

$$\wedge_{r_1 \cdots r_n} \phi(x) = O(|x|^{-n-1}), \qquad |x| \longrightarrow \infty$$

for $r_1 + \dots + r_n \geq r'$ and we have

$$\wedge \phi(x) \equiv 0$$

If we now take fixed values $r_1, \dots, r_n$, and fixed coordinates $x_0^{m+1}, \dots, x_0^n$ for which

(28) $$(x_0^{m+1})^2 + \dots + (x_0^n)^2 > R^2$$

and introduce the function

$$\psi(x^1, \ldots, x^m) = \Lambda_{r_1 \ldots r_n} \phi(x^1, \ldots, x^m, x_o^{m+1}, \ldots, x_o^n)$$

then the latter has the following properties: It is defined and analytic in E_m: $(x^1, \ldots, x^m)$, we have

(29) $$\Lambda_{q_1 \ldots q_m} \psi(x) = O(|x|^{-m-1}), \qquad |x| \longrightarrow \infty$$

for any $q_1 \geq 0, \ldots, q_m \geq 0$, and we have

(30) $$\Lambda \psi(x) \equiv 0$$

We claim that under these circumstances we have

$$\psi(x) \equiv 0$$

In fact, because of (29) we may introduce the Fourier transform

$$\chi(\alpha_j) = \int_{E^m} e^{i(\alpha_1 x^1 + \ldots + \alpha_m x^m)} \psi(x) dv_x$$

and it follows easily that the Fourier transform of $\Lambda\psi(x)$ is then

(31) $$\chi(\alpha_j) \quad \sum_{(q)} a_{q_1 \ldots q_m} (i\alpha_1)^{q_1} \ldots (i\alpha_m)^{q_m}$$

Now, (30) implies the vanishing of (31), and hence the vanishing of $\chi(\alpha)$, and by uniqueness of Fourier transforms the vanishing of $\psi(x)$. Therefore we have

(32) $$\Lambda_{r_1 \ldots r_n} \phi(x^1, \ldots, x^m, x^{m+1}, \ldots, x^n) = 0$$

for $(x^1, \ldots, x^m)$ in E_m and $(x^{m+1}, \ldots, x^n)$ in (28), but since the left side of (32) is analytic in (27), the relation (32) also holds identically in (27). And since this is true for all combinations $r_1, \ldots, r_n$ with $r_1 + \ldots + r_n = r'$ it follows that $\phi(x)$ is a polynomial of total degree $\leq r' - 1$, and thus is certainly continuable into all E_n.

§2. COORDINATE SPACES

We now take an arbitrary analytic coordinate space X_n: $(\xi^1, \ldots, \xi^n)$ and in it, immediately for $p = n - 1$, a form

$$G(\xi;\ x;\ f(\xi);\ d\xi) = \frac{1}{(n-1)!}\sum_{(\alpha)} A_{\alpha_1\cdots\alpha_{n-1}}\, d\xi^{\alpha_1} \cdots d\xi^{\alpha_{n-1}}$$

where

$$A_{\alpha_1\ldots\alpha_{n-1}} = \sum_{(\nu)} G^{\nu_1\ldots\nu_n}_{\alpha_1\cdots\alpha_{n-1}}(\xi;\ x)\wedge_{\nu_1\ldots\nu_n} f(\xi)$$

for $\xi \neq x$. The point x itself also ranges over X_n, and any two points ξ_0, x_0 $(\xi_0 \neq x_0)$ have non-overlapping coordinate neighborhoods for which the tensoroid components $G^{(\nu)}_{(\alpha)}(\xi;\ x)$ are defined and analytic in $(\xi;\ x)$.

We again introduce the requirement $d_\xi G = 0$, that is

$$\sum_{q=1}^{n} (-1)^q \frac{\partial}{\partial \xi^{\alpha_q}} A_{\alpha_1\cdots\alpha_{q-1}\alpha_{q+1}\cdots\alpha_n} = 0 \quad .$$

and it amounts to a tensoroid system of equations

$$\sum_{(\nu)} H^{(\nu)}_{(\alpha)}(\xi;\ x)\wedge_{\nu_1\ldots\nu_n} f(\xi) = 0$$

which we again abbreviate to

(33) $$\Delta_\xi f(\xi) = 0$$

and we again consider analytic functions in open sets U which satisfy this "elliptic equation" there. We wish to point out, however, that we are not retaining in any manner whatsoever the previous assumption that the variables ξ, x shall occur as the difference $\xi - x$ only, and thus the "constancy of coefficients" in the operator Δ is being entirely dispensed with.

We again introduce the integral

(34) $$g(x) = \int_{B_{n-1}} G(\xi;\ x;\ f(\xi);\ d\xi)$$

for a chain in U. It is again analytic and distributive, and we again have (14); and if B_{n-1} is the boundary of a small domain D^0, then for points of D^0 we have

$$g(x) = \sum_{(r)} c^{r_1\ldots r_n}(x)\wedge_{r_1\ldots r_n} f(x)$$

where the coefficients $c^{r_1 \ldots r_n}(x)$ are now functions of x.

We again call the integral a Green's formula if we have, in particular,

$$\oint_{B_{n-1}} G(\xi;\ x;\ f(\xi);\ d\xi) = g(x) \tag{35}$$

and theorem 4 is valid again.

We now take a second analytic coordinate space Y_m: $(y^1, \ldots, y^m)$, $(m \gtreqless n)$ which will be a space of "parameters" and we form the product space

$$Z_{n+m} = X_n \times Y_m \tag{36}$$

in the usual manner. Also, for any point set θ in Z_{n+m}, we will employ the representation by its layers in the X_n-space, thus

$$\theta = \left\{ y \in Y_m;\ x \in D(y) \right\} \tag{37}$$

where for each y, $D(y)$ is a point set in X_n, perhaps empty.

We now take a domain (37) in (36) and another such domain

$$\tilde{\theta} = \left\{ y \in Y_m;\ x \in \tilde{D}(y) \right\} \tag{38}$$

and we introduce the following

DEFINITION. We call the domain $\tilde{\theta}$ an enlargement of the domain θ if

(i) $\tilde{\theta} \supset \theta$ that is $\tilde{D}(y) \supset D(y)$ for every y,

(ii) for each (x,y) in $\tilde{\theta}$ there is in $D(y)$ a cycle $B_{n-1} = B_{n-1}(x,y)$ which bounds an n-complex $K_n(x,y)$ in $\tilde{D}(y)$, $B_{n-1} = bd(K_n)$, such that $x \in K_n(x,y)$.

(iii) for any (x',y') in $\tilde{\theta}$ there is a neighborhood $N = N(x',y')$ in $\tilde{\theta}$ such that for $(x,y) \in N$ the cycle $B_{n-1}(x',y') - B_{n-1}(x,y)$ bounds a chain $H_n(x,y)$ in $D(y)$, with $x \notin H_n(x,y)$, that is $B_{n-1}(x',y') \approx B_{n-1}(x,y)$ in $D(y) - x$; and finally

(iv) there exists a point (x_0, y_0) in θ and a neighborhood $N_0(x_0,y_0)$ in θ such that for any (x,y) in $N_0(x_0,y_0)$ the complex $K_n(x,y)$ of (ii) lies in $D(y)$ itself, and not only in the larger $\tilde{D}(y)$.

We note that condition (iv) is implied by the simpler though less general condition

(iv)' for (x,y) in $N_o(x_o,y_o)$ we have $\tilde{D}(y) = D(y)$.

Now, the leading statement is as follows:

THEOREM 6. If $f(x,y)$ is defined in a domain θ of Z_{n+m} and is analytic as a function of (x,y) there, and if for each y it is a solution of our elliptic equation

(39) $$\Delta_x f(x, y) = 0$$

then $f(x, y)$ has an analytic continuation into any given enlargement $\tilde{\theta}$ of θ.

REMARK. There are enlargements which in a certain sense are not enlargeable themselves. If, however, we take any two enlargements $\tilde{\theta}_1$, $\tilde{\theta}_2$ of the kind introduced and if the point sets $\tilde{\theta}_1 - \theta$, $\tilde{\theta}_2 - \theta$ have a non-vacuous intersection which has parts not connected with θ itself, then we do not claim that the two continuations will necessarily coincide there; it is not known to us what the actual situation is.

PROOF. For every (x,y) in $\tilde{\theta}$, we introduce the quantity

(40) $$g(x, y) = \int_{B_{n-1}(x, y)} G(\xi; x; f(\xi, y); d\xi).$$

Now, in a neighborhood $N(x', y')$ [defined on p. 15, property (iii)] we can replace the variable cycle $B_{n-1}(x, y)$ by the fixed cycle $B_{n-1}(x',y')$ and it now follows that $g(x, y)$ is analytic in $N(x', y')$ and hence everywhere in θ. But by property (iv) it coincides with $f(x, y)$ itself in a certain neighborhood N_o of θ, and by analytic continuation it coincides with $f(x, y)$ everywhere in θ, as claimed.

We now take in Z_{n+m} a closed point set which is the locus of a (finite or infinite) simplicial chain R_{n-1+m} and a neighborhood V_{n+m} of the latter and we also introduce the open set

(41) $$\theta_{n+m} = Z_{n+m} - R_{n-1+m}$$

We are also introducing the representatives

$$R_{n-1+m} = \left\{ y \in Y_n;\ \ x \in B_{n-1}(y) \right\}$$

$$V_{n+m} = \left\{ y \in Y_m; \quad x \in U_n \ (y) \right\}$$
$$\theta_{n+m} = \left\{ y \in Y_m; \quad x \in D_n \ (y) \right\}$$

and we are making the following assumptions.

(v) each $B_{n-1}(y)$ is the locus of a ("regular" or) "singular" cycle which we denote by the same symbol, and

(vi) for each (x', y') in θ_{n+m} there exists a neighborhood $N(x', y')$ and a cycle $B_{n-1}(x', y')$ in $U_n(y')$ such that for (x,y) in $N(x', y')$ the cycle $B_{n-1}(y) - B_{n-1}(x', y')$ bounds an n-dimensional complex in $U_n(y)$ which does not contain the point x.

THEOREM 7. Under the assumptions just made, if $f(x, y)$ is defined and analytic in V_{n+m} and satisfies (39) there, then the integral

$$g(x, y) = \int_{B_{n-1}(y)} G(\xi;\ x;\ f(\xi;\ y);\ d\xi) \tag{42}$$

defines an analytic function in θ_{n+m} having the following properties:

If we denote the (connected) components of θ_{n+m} by $\theta^1, \theta^2, \ldots,$ and denote the value of (42) in θ^a by $g^a(x, y)$, and if a "regular" simplex R^0_{n-1+m} of the chain R_{n-1+m} separates two components θ^a, θ^b; then the function $g^a(x, y)$ can be continued a certain distance across R^0 into θ^b, and $g^b(x, y)$ can be so continued into θ^a, and in the vicinity of R^0 we have the saltus relation

$$g^a(x, y) - g^b(x, y) = \pm f(x, y)$$

For the moment we call a component θ^a <u>non-bounded</u> if there is in Y_m some (small) neighborhood N_m such that for y in N_m the cross section $D^a_n(y)$ of $\theta^a_{n+m}(y)$ is the entire X_n. For these values y, the cycle $B_{n-1}(y)$ can be taken as null and thus $g^a(x, y) \equiv 0$. Hence the following consequences.

THEOREM 8. If in Theorem 7 all components θ^a are non-bounded then $f(x, y) \equiv 0$. If V_{n+m} is connected then the conclusion also holds if there are two non-bounded components θ^a, θ^b

having on their boundaries a joint simplex R^{o}_{n-1+m} separating them.

If there are altogether only two components θ^{1}, θ^{2}, and θ^{2} is non-bounded, then $f(x, y)$ can be continued from the intersection of V_{n+m} with θ^{1} into all of θ^{1}, and this is the most frequently occurring situation underlying Theorem 6.

II. STRONGLY ELLIPTIC SYSTEMS OF DIFFERENTIAL EQUATIONS

F. E. Browder

§1. INTRODUCTION

In a number of recent papers, we have presented a general theory of boundary-value problems for linear elliptic equations of arbitrary order and, more generally, for linear elliptic systems of differential equations ([3], [4], [5], [6], [7], [8]). It is the object of this paper to present a simple self-contained proof of the most basic results which we have obtained for the case of linear "strongly" elliptic systems of differential equations. These form a general subclass of the elliptic systems which contains single elliptic equations as well as such important special cases as the Laplace equation for exterior differential forms on Riemannian manifolds ([10]).

For the single elliptic equation, results similar to those of [3] have been announced by L. Gårding in [18]. The definition of strongly elliptic systems was given by M. I. Visik in [28]. Visik's theory of strongly elliptic systems presented in [29] has many points of contact with our results. Some major differences must be noted, however. The most important of these is that Visik obtains only weak solutions for his boundary-value problem (i.e., distributions in the sense of [24] or [26]) and establishes no analogue of the regularity theorem which is proved below. In particular, he obtains no results on fundamental solutions, Green's functions, or compactness and convergence theorems. In addition, Visik's abstract method rests upon results of Sobolev ([25], [26]) and Kondrashov ([21]) which are more complicated in character than the techniques which we employ. Morrey in [22] has also discussed strongly elliptic systems of second-order equations in two independent variables.

Our basic regularity theorem asserting that a weak solution of our equations is essentially a strict solution in the classical sense was established for a single elliptic equation in [5] and for a general elliptic system in [8] using an extension of the method of F. John ([19]) for the construction of a sufficiently differentiable fundamental solution in the small. In this paper, however, for the sake of directness and simplicity, we prove this theorem for strongly elliptic systems using the ideas and techniques of the Friedrichs mollifier method ([12], [13], [14]). Though it yields weaker estimates and is definitely restricted to the strongly elliptic case, this method is more closely related to our abstract approach than is the fundamental solution method. Friedrichs has recently presented such a proof in [15]. In [20], John has announced the construction, for a single linear elliptic equation, of a proof of similar type based on the method of spherical means discussed in Chapter IV of [19].

§2 presents the detailed formulation of the theorems which are proved in this paper. §3 contains the proof of auxiliary lemmas to be used in the later sections. In §4 the semi-boundedness of the general strongly elliptic linear system of differential equations is established. §5 is devoted to the proof of our basic regularity theorem for weak solutions of a strongly elliptic system of equations. In §6, the basic theorems concerning the Dirichlet problem are established including the Fredholm alternative, the discreteness of eigenvalues and finite dimensionality of eigenspaces, as well as the completeness of the eigenfunctions of self-adjoint systems. (For the proof of the completeness of the eigenfunctions of elliptic equations and strongly elliptic systems which are not necessarily self-adjoint, cf. [7] and [8]). §7 concludes the discussion with the proof of the existence and regularity of the Green's function for domains on which the Dirichlet problem has a unique solution.

§2. FORMULATION OF THEOREMS

Let D be a bounded domain in Euclidean n-space E^n. (Some partial extensions of our results to unbounded domains are given in [4] and [8]). We shall consider several families of complex-valued functions on D. If j is a non-negative integer, $C^j(D) = \{f | f$ and all its partial derivatives of order $\leq j$ are defined and continuous on $D\}$, $C^j(\overline{D}) = \{f | f$ and all its derivatives of order $\leq j$ are uniformly continuous on $D\}$; $L_2(D)$ is the Hilbert space of complex-valued square-summable functions on D. $C_c^\infty(D) = \{\phi | \phi$ and all its partial derivatives are defined and continuous on D, ϕ vanishes outside a compact subset of $D\}$. If $\phi \in C_c^\infty(D)$, $S(\phi)$ is the closure of the set $\{x | \phi(x) \neq 0\}$.

We shall consider r-vector functions, i.e., vector functions with r components $u = (u_1, \ldots, u_r)$; the i-th component u_i of u is a complex-valued function defined on D. $C^{j,r}(D)$, $C^{j,r}(\overline{D})$, $L_{2,r}(D)$, $C_c^{\infty,r}(D)$ are defined as the families of r-vector functions u such that for each i, u_i belongs to $C^j(D)$, $C^j(\overline{D})$, $L_2(D)$, $C_c^\infty(D)$ respectively.

A system K of differential equations on D of order $2m$ and rank r has the form

$$K_i(u) = \sum_{j=1}^{r} \sum_{\substack{k_1,\ldots,k_s=1 \\ s \leq 2m}}^{n} a_{k_1\ldots k_s;i,j}(x) \frac{\partial^s u_j}{\partial x_{k_1}\ldots\partial x_{k_s}} \tag{2.1}$$

$$= v_i \qquad (i = 1, \ldots, r)$$

where the indices $k_1, \ldots, k_s$ range independently from 1 to n while for each set of indices $a_{k_1\ldots k_s;i,j} \in C^s(D)$. The system of differential operators K transforms $u \in C^{2m,r}(D)$ into $v \in C^{0,r}(D)$. For the sake of simplicity we assume the coefficients real. For each system K, we may define a r by r characteristic matrix $A(x, \xi)$, defined for x in D and every real n-vector $\xi = (\xi_1, \ldots, \xi_n)$ and depending only upon the highest order terms of the system K:

$$A(x, \xi) = (a_{ij}(x, \xi)) \tag{2.2}$$

$$= \left(\sum_{k_1\ldots k_{2m}} a_{k_1\ldots k_{2m};i,j}(x)\xi_{k_1}\ldots\xi_{k_{2m}} \right)$$

DEFINITIONS:

$\mathcal{E}$) K is said to be elliptic at x if $A(x, \xi)$ is non-singular for every $\xi \neq 0$.

$S\mathcal{E}_1$) K is said to be strongly elliptic at x in D if $A(x, \xi) + A^t(x, \xi)$ is positive definite for every $\xi \neq 0$, where A^t is the transpose of A. An equivalent formulation is the following:

$S\mathcal{E}_2$) Given $x \in D$, there exists $\rho > 0$ such that for every real n-vector ξ and r-vector η,

$$\sum_{i,j=1}^{r} \sum_{k_1\ldots k_{2m}=1}^{n} a_{k_1\ldots k_{2m};i,j}(x)\xi_{k_1}\ldots\xi_{k_{2m}}\eta_i\eta_j \tag{2.3}$$

$$\geq \rho \left(\sum_{i=1}^{n} \xi_i^{2m} \right) \left(\sum_{j=1}^{r} \eta_j^2 \right)$$

K will be said to be uniformly strongly elliptic on D if there exists $\rho > 0$ for which $S\mathcal{E}_2$ is satisfied for all x in D.

For a single equation $(r = 1)$, $S\mathcal{E}_2$ reduces to the classical criterion of ellipticity. As a consequence the theory of strongly elliptic systems includes the theory of the linear elliptic differential equation of arbitrary order. From the definitions, it follows by a formal argument that the strongly elliptic systems are a proper sub-class of the elliptic systems. It has been shown by an example in [30] that such results of the theory of strongly elliptic systems as the discreteness of eigenvalues in the Dirichlet problem are not true for all elliptic systems.

For $\phi \in C_c^{\infty,r}(D)$, we define

$$\|\phi\|_m^2 = \sum_{j=1}^{r} \sum_{k_1 \ldots k_m = 1}^{n} \int_D \left| \frac{\partial^m \phi_j}{\partial x_{k_1} \ldots \partial x_{k_m}} \right|^2 dx$$

(Integration is taken with respect to Lebesgue n-measure.)

For $u, v \in L_{2,r}(D)$,

$$(u \cdot v) = \sum_{j=1}^{r} \int_D u_j v_j^* \, dx$$

(z^* = complex conjugate of z).

The basic property of strongly elliptic systems which is not shared by the general class of elliptic systems is the semi-boundedness property expressed in the following theorem:

THEOREM 1. Suppose D is a bounded domain in E^n, K a system of differential operators which is uniformly strongly elliptic on D. Suppose that each of the coefficients $a_{k_1 \ldots k_s; i, j} \in C^{\gamma(s)}(\overline{D})$, where $\gamma(s) = \max\{0, s - m\}$. Then there exist $\rho_1 > 0$, $k_0 \geq 0$ such that for all $\phi \in C_c^{\infty,r}(D)$,

$$\operatorname{Re}\{(-1)^m K(\phi) \cdot \phi\} \geq \rho_1 \|\phi\|_m^2 - k_0(\phi \cdot \phi)$$

If K is strongly elliptic on D, the conditions of Theorem 1 will be satisfied on all subdomains which are contained in compact subsets of D.

Theorem 1 enables us to translate boundary value problems into abstract problems concerning the existence of solutions of linear functional equations in a Hilbert space. The essential tool that guarantees that the translation is reversible and unambiguous is furnished by Theorem 2, our basic regularity theorem for weak solutions of the system K.

We may associate with each system of the form (2.1) the adjoint system $\overline{K}$ having the form:

$$\overline{K}_j(u) = \sum_{i=1}^{r} \sum_{\substack{k_1 \ldots k_s = 1 \\ s \leq 2m}}^{n} (-1)^s \frac{\partial^s}{\partial x_{k_1} \ldots \partial x_{k_s}} [a_{k_1 \ldots k_s; i, j}(x) u_i] \tag{2.4}$$

$$(j = 1, \ldots, r)$$

It is shown in an elementary way in §3 that for $\phi \in C_c^{\infty, r}(D)$ and $u \in C^{2m, r}(D)$, $Ku \cdot \phi = u \cdot \overline{K}\phi$. If $Ku = \gamma$, it follows that $u \cdot \overline{K}\phi = \gamma \cdot \phi$ for all $\phi \in C_c^{\infty, r}(D)$. Theorem 2 allows us to reverse the direction of this argument.

If the components of u are square summable on compact subsets of D, u is said to be m-times strongly differentiable on D if on every compact subset C of D, u is the limit in $L_{2,r}(C)$ of a sequence $\{\phi_{,t}\}$, $\phi_{,t} \in C^{m,r}(\overline{C})$ for which

$$\frac{\partial^{m_1} \phi_{,t}}{\partial x_{k_1} \ldots \partial x_{k_{m_1}}}$$

is a Cauchy sequence in $L_{2,r}(C)$ for every set of indices $(k_1, \ldots, k_{m_1})$, $m_1 \leq m$.

THEOREM 2. Let D be a domain in E^n, K a strongly elliptic system of differential operators on D.

(a) Suppose that for the coefficients in K, $a_{k_1 \ldots k_s; i, j} \in C^{s+M}(D)$ where

$$M = \left[\frac{n+1}{2}\right], \quad \gamma \in C^{M,r}(D) \cap L_{2,r}(D)$$

Suppose that u is an r-vector function on D, m-times strongly differentiable in D, and satisfying the equation $u \cdot \overline{K}\phi = \gamma \cdot \phi$ for all $\phi \in C_c^{\infty, r}(D)$. Then $u = v$ almost everywhere in D for a suitable function $v \in C^{2m,r}(D)$ for which $Kv = \gamma$.

(b) If the coefficients $a_{k_1 \dots k_s; i,j}$ lie in $C^{s+M+t}(D)$, $\gamma \in C^{M+t,r}(D)$ for $t > 0$, then v lies in $C^{2m+t,r}(D)$.

Strong differentiability of u is introduced as a simplifying hypothesis. We have shown in [5] and [8] that Theorem 2 holds without this condition, and have established convergence and compactness theorems of the Harnack type for solutions of elliptic systems.

We shall restrict our attention to strongly elliptic systems of equations $Ku = \gamma$ with suitably differentiable coefficients and inhomogeneous parts, i.e., such that the differentiability conditions in (a) of Theorem 2 are satisfied. For such systems, the main conclusion of Theorem 2 will be valid.

In order to formulate the Dirichlet problem in its general form, it is necessary to introduce the definition of the variational boundary conditions. If $m \geq 1$, we define $H_{m,r}(D)$, the space of functions vanishing on the boundary of D in the variational sense together with all their derivatives of order less than m, as the set of those $u \in L_{2,r}(D)$ such that a sequence $\{\phi_{,s}\}$ may be chosen from $C_c^{\infty,r}(D)$ converging to u in $L_{2,r}(D)$ and such that for every set of m indices $k_1, \dots, k_m$,

$$\left\{ \frac{\partial^m \phi_{,s}}{\partial x_{k_1} \dots \partial x_{k_m}} \right\}$$

is a Cauchy sequence in $L_{2,r}(D)$.

Given $g \in C^{m,r}(\overline{D})$ and $\gamma \in C^{M,r}(D)$, the Dirichlet problem for the strongly elliptic system K on D asks for $u \in C^{2m,r}(D)$ for which $Ku = \gamma$ while $u - g$ lies in $H_{m,r}(D)$, i.e., $u - g$ and all its derivatives of order less than m vanish on the boundary of D in the variational sense.

THEOREM 3. Let K be a uniformly strongly elliptic system of differential operators on the bounded domain D of E with suitably differentiable coefficients. Suppose that for each of the coefficients in K, $a_{k_1 \dots k_s; i,j} \in C^{\gamma(s)}(\overline{D})$. Then:

(a) The solution of the Dirichlet problem for $Ku = \gamma$ with boundary function g exists if there is at most one such solution.

(b) The solution of the Dirichlet problem for $Ku = \gamma$ with zero boundary function exists if and only if $\gamma \cdot \theta = 0$ for every solution of the homogeneous adjoint system $\overline{K}\theta = 0$ with zero boundary function. The linear dimension of the null solutions for $\overline{K}$ is equal to that for K.

(c) There exists $d > 0$ such that for every sub-region $D_1 \subset D$ with $\text{diam}(D_1) < d$, the Dirichlet problem for K on D_1 has a unique solution for every inhomogeneous part γ and boundary function g.

The relation of the variational boundary conditions to the classical Dirichlet boundary conditions is discussed in detail in [8]. In [5], and [8], we have shown that for the general linear elliptic system in two independent variables, the solution of the Dirichlet problem assumes the full set of boundary values in the pointwise sense for all domains D with locally connected boundary with no isolated points.

If $u \in C^{2m,r}(D)$, u is said to be an eigenfunction of K on D with eigenvalue λ if $Ku = \lambda u$ while $u \in H_{m,r}(D)$.

THEOREM 4. Let K be a uniformly strongly elliptic system of differential operators on the bounded domain D of E^n with suitably differentiable coefficients. Suppose that for each of the coefficients of K, $a_{k_1 \ldots k_s; i,j} \in C^{\gamma(s)}(\overline{D})$. Then:

(a) The eigenfunctions of $(-1)^m K$ are discrete with their real parts uniformly bounded from below, while the linear dimension of the set of eigenfunctions corresponding to any single eigenvalue is finite.

(b) If K is a self-adjoint system, $(K = \overline{K})$, then the eigenfunctions of K on D are complete in $L_{2,r}(D)$ and $H_{m,r}(D)$.

The r by r matrix function $E(x, z) = (e_{ij}(x, z))$ defined for x, z in D, $x \neq z$, is said to be a fundamental solution for K on D if for all x in D,

$$\phi_i(x) = \sum_{j=1}^{r} \int_D e_{ij}(x, z)\overline{K}\phi_j(z)\, dz \quad (i = 1, \ldots, r) \qquad \phi \in C_c^{\infty,r}(D)$$

The matrix function $G(x, z) = (g_{ij}(x, z))$, defined for x, z in D, $x \neq z$, is said to be a Green's function for K on D if

$$u_j(z) = \sum_{i=1}^{r} \int_D g_{ij}(x, z)\gamma_i(x)\, dx$$

for $\gamma \in C^{M,r}(D) \cap L_{2,r}(D)$ and u is the solution of the Dirichlet problem for $Ku = \gamma$ with zero boundary values.

THEOREM 5. Let K be a strongly elliptic system on the bounded domain D of E. Suppose that the coefficients of K satisfy the differentiability conditions of the hypothesis of Theorem 3 as well as (b) of Theorem 2 for $t = \gamma[M - m]$ and that the Dirichlet problem for K on D has no non-trivial null solutions. Then there exists a (unique) Green's matrix $G(x, z)$ for K on D. The components of $G(x, z)$ are 2m-times continuously differentiable in x and z separately for $x \neq z$. $G(x, z)$ is a fundamental solution for K on D while $H(x, z) = G^t(z, x)$ is a fundamental solution for $\overline{K}$ on D. If the conditions of order t in (b) of Theorem 2 are satisfied, then $G(x, z)$ is $(2m + t)$-times continuously differentiable in each variable for $x \neq z$.

§3. AUXILIARY LEMMAS

We designate by $\overline{D}$ the closure of D in E^n; $d(x, y)$, the Euclidean distance between x and y in E^n, and by $d(x, A) = \underset{y \in A}{\text{g.l.b.}}\, d(x, y)$ for x in E^n, $A \subset E^n$. If $\epsilon > 0$, $D_\epsilon = \{x \mid d(x, E^n - D) > \epsilon\}$, $N_\epsilon(D) = \{x \mid d(x, D) < \epsilon\}$.

We choose the non-negative valued function $j(t)$ defined and infinitely differentiable on the real line satisfying the conditions

(a) $j(t) = 0$ for $t \geq 1$, $j(t) \geq 0$ for all t

(b) $$\int_{-\infty}^{+\infty} j(t)\, dt = 1$$

For each $\epsilon > 0$, we define the function $j_\epsilon(x)$ in E^n by

(3.1) $$j_\epsilon(x) = \epsilon^{-n} \prod_{i=1}^{n} j(x_i/\epsilon)$$

If $u \in L_{2,r}(D)$, $x^1 \in D$, let

(3.2) $$J_\epsilon(u)_i(x^1) = \int_D j_\epsilon(x - x^1)u_i(x)\,dx \quad (i = 1, \ldots, r)$$

If all the components of u are set equal to zero outside of D, we have

(3.3) $$J_\epsilon(u)_i(x^1) = \int_{E^n} j_\epsilon(x - x^1)u_i(x)\,dx, \quad x^1 \in D$$

A fundamental property of this operator is that

$$\frac{\partial}{\partial x_j^1} J_\epsilon(u)_i = -\int_D \frac{\partial j_\epsilon(x - x^1)}{\partial x_j} u_i(x)\,dx$$

Clearly $J_\epsilon(u) \in C^{\infty,r}(\bar{D}) \cap L_{2,r}(D)$. If each of the components of u vanishes almost everywhere outside a compact subset of D_ϵ, then $J_\epsilon(u) \in C_c^{\infty,r}(D)$. Further, if $u_i = 1$ on $D_1 \subset D$, then $J_\epsilon(u)_i(x^1) = 1$ on $(D_1)_\epsilon$. Let $u_\delta(x)$ be an infinitely often differentiable function, $0 \le u_\delta \le 1$ and which equals 1 for x in $D_{3\delta/4}$, 0 outside $D_{\delta/2}$. Define $h_\delta = J_{\delta/4}(u_\delta)$. Then $h_\delta \in C_c^\infty(D)$ and $h_\delta(x) = 1$ for x in D_δ.

If $u \in C^{2m,r}(D)$, $\phi \in C_c^{\infty,r}(D)$, $\delta = d(S(\phi), E^n - D)$, $h_\delta u$ and ϕ may be extended to r-vector functions on any n-cube N containing D in its interior by setting their components equal to zero outside D. By integration by parts it follows that for $s \le 2m$,

(3.4) $$\begin{aligned}
&\int_D a_{k_1\ldots k_s;i,j}(x) \frac{\partial^s u_j}{\partial x_{k_1}\ldots\partial x_{k_s}} \phi_i^*\,dx \\
&= \int_D a_{k_1\ldots k_s;i,j}(x) \frac{\partial^s (h_\delta u_j)}{\partial x_{k_1}\ldots\partial x_{k_s}} \phi_i^*\,dx \\
&= (-1)^s \int_D h_\delta u_j \frac{\partial^s\left(a_{k_1\ldots k_s;i,j}(x)\phi_i^*\right)}{\partial x_{k_1}\ldots\partial x_{k_s}}\,dx \\
&= (-1)^s \int_D u_j \frac{\partial^s\left(a_{k_1\ldots k_s;i,j}(x)\phi_i^*\right)}{\partial x_{k_1}\ldots\partial x_{k_s}}\,dx
\end{aligned}$$

It follows by summation that

$$Ku \cdot \phi = u \cdot \bar{K}\phi \tag{3.5}$$

For $\phi, \Psi \in C_c^{\infty, r}(D)$, we define the s-inner product and s-norm by the formulae:

$$(\phi, \Psi)_s = \sum_{i=1}^{r} \int_D \sum_{k_1 \ldots k_s = 1}^{n} \left\{ \frac{\partial^s \phi_i}{\partial x_{k_1} \ldots \partial x_{k_s}} \frac{\partial^s \Psi_i^*}{\partial x_{k_1} \ldots \partial x_{k_s}} \right\} dx$$

(3.6)

$$\|\phi\|_s = (\phi, \phi)_s^{1/2}$$

LEMMA 1. Let D be a bounded domain in E^n of diameter d. Then for all $s < k$,

$$\|\phi\|_s \leq d^{k-s}\|\phi\|_k, \quad \phi \in C_c^{\infty, r}(D) \tag{3.7}$$

PROOF. It suffices to prove the lemma for a real-valued $\phi \in C_c^{\infty}(D)$. Since D is of diameter less than d, it is contained in a n-cube C of side length d of the form $C = \{x | a_i \leq x_i \leq b_i;\ b_i - a_i = d \ (i = 1, \ldots, n)\}$. By setting its components equal to 0 outside of D, ϕ may be extended to an element of $C_c^{\infty}(C)$.

For $x \in D$, $\phi \in C_c^{\infty}(C)$,

$$\phi(x_1, \ldots, x_n) = \int_{a_1}^{x_1} \frac{\partial \phi(t, x_2, \ldots, x_n)}{\partial x_1} dt \tag{3.8}$$

By the Schwarz inequality,

$$|\phi(x_1, \ldots, x_n)|^2 \leq d \int_{a_1}^{b_1} \left| \frac{\partial \phi(t, x_2, \ldots, x_n)}{\partial x_1} \right|^2 dt$$

Integrating with respect to $x_1, \ldots, x_n$ over their limits on C,

$$\|\phi\|_0^2 \leq d^2 \|\phi\|_1^2 \tag{3.9}$$

Applying (3.9) to the various t-th order derivatives of ϕ and summing,

(3.10) $$\|\phi\|_t^2 \leq d^2\|\phi\|_{t+1}^2, \quad \text{i.e.,} \quad \|\phi\|_t \leq d\,\|\phi\|_{t+1}$$

Applying (3.10) for $s \leq t < k$, $k - s$ times, the Lemma follows. Q.E.D.

Since by Lemma 1, $\|\phi\|_s = 0$ only if $\|\phi\|_0 = 0$, $C_c^{\infty,r}(D)$ may be completed with respect to the s-inner product to an abstract Hilbert space $H^{s,r}(D)$, consisting of equivalence classes of Cauchy sequences from $C_c^{\infty,r}(D)$ with respect to the s-norm. By Lemma 1, however, each such sequence is also a Cauchy sequence in $L_{2,r}(D)$ while equivalent Cauchy sequences with respect to the s-norm are equivalent with respect to the 0-norm. It follows that $H^{s,r}(D)$ may be mapped linearly and continuously into $L_{2,r}(D)$ by an extension of the identification mapping on $C_c^{\infty,r}(D)$. The image of $H^{s,r}(D)$ under this mapping is easily seen to be $H_{s,r}(D)$ as defined in §2. Moreover, this mapping is a one-one mapping of $H^{s,r}(D)$ into $L_{2,r}(D)$. For suppose that $\{\phi_{,t}\}$ is a Cauchy sequence from $C_c^{\infty,r}(D)$ converging to $h \in H^{s,r}(D)$ in the s-norm while we have $\{\phi_{,t}\}$ converging to 0 in the 0-norm. By the continuity of norm

$$\|h\|_s^2 = \sum_{i=1}^{r} \sum_{k_1\ldots k_s=1}^{n} \|h_{k_1\ldots k_s,i}\|_0^2$$

where $h_{k_1\ldots k_s,i}$ is the limit in $L_2(D)$ of the Cauchy sequence

$$\left\{ \frac{\partial^s (\phi_{,t})_i}{\partial x_{k_1}\ldots\partial x_{k_s}} \right\}$$

Let ψ be in $C_c^{\infty}(D)$. Then

$$\int_D h_{k_1\ldots k_s,i}(x)\psi^*(x)\,dx$$

(3.11) $$= \lim_{t\to\infty} \int_D \frac{\partial^s (\phi_{,t})_i}{\partial x_{k_1}\ldots\partial x_{k_s}} \psi^*(x)\,dx$$

$$= \lim_{t\to\infty} (-1)^s \int_D (\phi_{,t})_i \frac{\partial^s \psi^*}{\partial x_{k_1}\ldots\partial x_{k_s}}\,dx$$

Since $\{\phi_{,t}\}$ converges to 0 in the 0-norm, the term on the right in

(3.11) is zero. Thus $h_{k_1 \ldots k_s, i}$ is orthogonal to $C_c^{\infty}(D)$ in $L_2(D)$ and, since it will be shown in Lemma 2 that $C_c^{\infty}(D)$ is dense in $L_2(D)$, $h_{k_1 \ldots k_s, i} = 0$ almost everywhere in D. Thus $\|h\|_s = 0$ and h must be the zero element of $H^{s,r}(D)$.

Since $H^{s,r}(D)$ is in one-to-one correspondence with $H_{s,r}(D)$, we shall impose the Hilbert space structure of the former on the latter and consider $H_{s,r}(D)$ with the s-inner product as the completion of $C_c^{\infty,r}(D)$ with respect to the s-norm.

If D_0 is a subdomain of D, we shall designate by $\|u\|_{s,D_0}$ the s-norm of u on D_0, while reserving the s-norm without extra subscript for D itself.

LEMMA 2. (a) Suppose $u \in L_{2,r}(D)$. Then $J_\epsilon(u)$ converges to u in the 0-norm on D as $\epsilon \longrightarrow 0$. Further $C_c^{\infty,r}(D)$ is dense in $L_{2,r}(D)$.

(b) Suppose u is s-times strongly differentiable in D. Then $J_\epsilon(u) - J_\eta(u)$ converges to 0 in the s-norm on every compact subset of D as $\epsilon, \eta \longrightarrow 0$.

(c) Let $a \in C^{s+1}(D)$, $u \in L_2(D)$. Suppose u is s-times strongly differentiable on D. If

$$R_\epsilon(u) = \int_D \left[a(x) - a(x^1)\right] \frac{\partial^{s+1} j_\epsilon(x-x^1)}{\partial x_{k_1} \ldots \partial x_{k_{s+1}}} u(x)\, dx$$

Then $R_\epsilon(u) - R_\eta(u)$ converges to 0 in the 0-norm on every compact subset of D as $\epsilon, \eta \longrightarrow 0$.

PROOF. We may, without loss of generality, assume that $r = 1$.
(a) If $u \in L_2(D)$, $\|J_\epsilon(u)\|_0 \leq \|u\|_0$; for,

$$\int_D \left| \int_D j_\epsilon(x - x^1) u(x)\, dx \right|^2 dx^1$$

$$\leq \int_D \left| \int_D j_\epsilon(x - x^1)\, dx \right| \cdot \int_D j_\epsilon(x - x^1) |u(x)|^2\, dx\, dx^1 \tag{3.12}$$

$$= \int_D \int_D j_\epsilon(x - x^1) |u(x)|^2\, dx^1\, dx = \|u\|_0^2$$

since

$$\int_D J_\epsilon(x - x^1)\, dx = 1$$

If $\delta > 0$ is given, there exists $v \in C_c^0(D)$ such that

$$\|u - v\|_0 \leq \frac{\delta}{3}$$

By the preceding remark $\|J_\epsilon(u) - J_\epsilon(v)\|_0 = \|J_\epsilon(u - v)\|_0 \leq \|u - v\|_0 \leq \frac{\delta}{3}$. Since v is uniformly continuous on D, its oscillation

$$Os_\epsilon(v) = \sup\{\, |v(x) - v(y)| : x, y \in D, |x - y| \leq \epsilon \,\}$$

approaches 0 as $\epsilon \longrightarrow 0$. For $x^1 \in D$, $\epsilon < d(E^n - D, S(v))$,

$$(3.13) \qquad \begin{aligned} &|J_\epsilon(v)(x^1) - v(x^1)| \\ &= \left| \int_D J_\epsilon(x - x^1)\left\{ v(x) - v(x^1)\right\} dx \right| \leq Os_\epsilon(v) \end{aligned}$$

If ϵ is sufficiently small, $\|J_\epsilon(v) - v\|_0$ can be made less than $\frac{\delta}{3}$. Then

$$(3.14) \qquad \begin{aligned} &\|J_\epsilon(u) - u\|_0 \leq \|J_\epsilon(u) - J_\epsilon(v)\|_0 \\ &+ \|J_\epsilon(v) - v\|_0 + \|v - u\|_0 < \delta \end{aligned}$$

The remainder of (a) follows if we note that $J_\epsilon(v) \in C_c^\infty(D)$.

(b) Let $\rho > 0$. Suppose $u \in L_2(D)$ and $\{\phi_{,t}\}$ is a sequence from $C^s(\overline{D}_\rho)$ converging to u in the 0-norm in D_ρ while $\{\phi_{,t}\}$ is a Cauchy sequence in the s-norm on D_ρ. Suppose that $c(x) \in C^s(\bar{D})$. We shall show somewhat more generally that, for any $c(x)$ in $C^s(D)$ and $S_\epsilon(u)$ defined by

$$S_\epsilon(u) = \int_D c(x) J_\epsilon(x - x^1) u(x)\, dx$$

the $S_\epsilon(u)$ form a Cauchy sequence in the s-norm on $D_{2\rho}$ as $\epsilon \longrightarrow 0$. (To prove (b), it suffices to take $c = 1$; the more general situation is used in the proof of (c)). Since $\{\phi_{,t}\}$ is a Cauchy sequence with respect to the s-norm on D_ρ, for every set of indices $(k_1, \ldots, k_s)$, there exists $h_{k_1 \ldots k_s}$ in $L_2(D_\rho)$ to which

$$\frac{\partial^s (c\phi_{,t})}{\partial x_{k_1} \dots \partial x_{k_s}}$$

converges in the 0-norm on D_ρ as $t \longrightarrow \infty$. It suffices for our purpose to show that

$$\frac{\partial^s [S_\epsilon(u) - S_\eta(u)]}{\partial x^1_{k_1} \dots \partial x^1_{k_s}}$$

converges to zero in the 0-norm on $D_{2\rho}$ as $\epsilon, \eta \longrightarrow 0$. We note by integration by parts that, for $\epsilon < \rho$

$$\left\| \frac{\partial^s S_\epsilon(\phi_{,t})}{\partial x^1_{k_1} \dots \partial x^1_{k_s}} - (-1)^s J_\epsilon (h_{k_1 \dots k_s}) \right\|_{0,D_{2\rho}}$$

$$(3.15) \qquad = \left\| J_\epsilon \left(\frac{\partial^s (c\phi_{,t})}{\partial x_{k_1} \dots \partial x_{k_s}} - h_{k_1 \dots k_s} \right) \right\|_{0,D_{2\rho}}$$

$$\leq \left\| \frac{\partial^s (c\phi_{,t})}{\partial x_{k_1} \dots \partial x_{k_s}} - h_{k_1 \dots k_s} \right\|_{0,D_{2\rho}}$$

and the last term approaches 0 as $t \longrightarrow 0$ independently of ϵ. By (a), given $\delta > 0$, there exists $\epsilon_0 > 0$ such that for $\epsilon, \eta < \epsilon_0$, $\|J_\epsilon (h_{k_1 \dots k_s}) - J_\eta (h_{k_1 \dots k_s})\|_{0,D_{2\rho}} < \frac{\delta}{5}$. If ϵ, η are fixed, we may choose t so large that

$$\left\| \frac{\partial^s S_\epsilon(\phi_{,t})}{\partial x^1_{k_1} \ldots \partial x^1_{k_s}} - (-1)^s J_\epsilon(h_{k_1 \ldots k_s}) \right\|_{0,D_{2\rho}}$$

$$\left\| \frac{\partial^s S_\eta(\phi_{,t})}{\partial x^1_{k_1} \ldots \partial x^1_{k_s}} - (-1)^s J_\eta(h_{k_1 \ldots k_s}) \right\|_{0,D_{2\rho}}$$

(3.16)

$$\left\| \frac{\partial^s [S_\epsilon(u) - S_\epsilon(\phi_{,t})]}{\partial x^1_{k_1} \ldots \partial x^1_{k_s}} \right\|_{o,D_{2\rho}}$$

$$\left\| \frac{\partial^s [S_\eta(u) - S_\eta(\phi_{,t})]}{\partial x^1_{k_1} \ldots \partial x^1_{k_s}} \right\|_{o,D_{2\rho}}$$

are all less than $\frac{\delta}{5}$. Then by the triangle inequality for the O-norm on $D_{2\rho}$,

$$\left\| \frac{\partial^s [S_\epsilon(u) - S_\eta(u)]}{\partial x^1_{k_1} \ldots \partial x^1_{k_s}} \right\|_{o,D_{2\rho}} < \delta \quad \text{for} \quad \epsilon,\ \eta < \epsilon_0$$

Choosing $c = 1$, (b) follows.

(c) $R_\epsilon(u)$ may be written in the form $R'_\epsilon(u) + T_\epsilon(u)$, where

$$R'_\epsilon(u) = \int_D \frac{\partial^{s+1}}{\partial x_{k_1} \ldots \partial x_{k_{s+1}}} \left\{ [a(x) - a(x^1)] j_\epsilon(x - x^1) \right\} u(x)\, dx$$

$$T_\epsilon(u) = \sum_{\substack{m_1 \ldots m_j \\ j \le s}} \frac{\partial_j}{\partial x^1_{m_1} \ldots \partial x^1_{m_j}} \int_D d_j(x)\ j_\epsilon(x - x^1) u(x)\, dx$$

with $d_j \in C^j(D)$. By the proof of (b) with d_j playing the role of $c(x)$, $T_\epsilon(u) - T_\eta(u)$ converges to zero in the O-norm on every compact subset of

D. It suffices to prove that $R'_\epsilon(u) - R'_\eta(u)$ converges to 0 in the 0-norm on every D_ρ for $\rho > 0$.

Let

$$M_\epsilon(u) = \int_D \frac{\partial}{\partial x_{k_1}} \left\{ [a(x) - a(x^1)] j_\epsilon(x - x^1) \right\} u(x)\, dx$$

Let $k'_1 = \sup\left\{ |\operatorname{grad} a(x)| : x \in D_\rho \right\}$, $k'_2 = \int_{-\infty}^{\infty} |j'(t)|\, dt$. It follows from the definition of $j_\epsilon(x)$ and the mean value theorem that

$$\int_D \left| \frac{\partial}{\partial x_{k_1}} \left\{ [a(x) - a(x^1)] j_\epsilon(x - x^1) \right\} \right| dx \le k'_1 k'_2 + k'_1$$

It follows by the argument of (3.12) that

$$\left\| \int_D \frac{\partial}{\partial x_{k_1}} \left\{ [a(x) - a(x^1)] j_\epsilon(x - x^1) \right\} u(x)\, dx \right\|_{o,D_{2\rho}}$$

$$\le (k'_1 k'_2 + k'_1) \|u\|_{o,D_\rho}$$

for $\epsilon = \rho$. We note that

$$\int_D \frac{\partial}{\partial x_{k_1}} \left\{ [a(x) - a(x^1)] j_\epsilon(x - x^1) \right\} dx = 0$$

As a first consequence of this fact, we prove that $M_\epsilon(u) \longrightarrow 0$ in the 0-norm on $D_{2\rho}$ as $\epsilon \longrightarrow 0$. For, given $\delta > 0$, let $v \in C_c^o(D_\rho)$ be such that

$$\|u - v\|_{o,D_{2\rho}} < \frac{\delta}{2(k'_1 k'_2 + k'_1)}$$

Then

$$|M_\epsilon(v)(x^1)| = \left| \int_D \frac{\partial}{\partial x_{k_1}} \left\{ [a(x) - a(x^1)] j_\epsilon(x - x^1) \right\} [v(x) - v(x^1)] \right| dx$$

$$\le (k'_1 k'_2 + k'_1) \operatorname{Os}_\epsilon(v) \longrightarrow 0 \quad \text{as} \quad \epsilon \longrightarrow 0$$

If ϵ is sufficiently small,

$$\|M_\epsilon(u)\|_{o,D_{2\rho}} \leq \|M_\epsilon(u) - M_\epsilon(v)\|_{o,D_{2\rho}} + \|M_\epsilon(v)\|_{o,D_\rho} < \delta$$

Let $\{\phi_{,t}\}$ be a sequence from $C^\infty(\overline{D}_\rho)$ converging to u in the 0-norm on D_ρ and forming a Cauchy sequence in the s-norm on D_ρ. Then

$$\|R'_\epsilon(u) - R'_\eta(u)\|_{o,D_{2\rho}}$$

$$\leq \|R'_\epsilon(u) - R'_\epsilon(\phi_{,t})\|_{o,D_{2\rho}} + \|R'_\eta(u) - R'_\eta(\phi_{,t})\|_{o,D_{2\rho}}$$
$$+ \|R'_\epsilon(\phi_{,t}) - R'_\eta(\phi_{,t})\|_{o,D_{2\rho}}$$

(3.17)

$$R'_\epsilon(u)(x^1) - R'_\epsilon(\phi_{,t})(x^1)$$
$$= \int_D \frac{\partial^{s+1}}{\partial x_{k_1} \cdots \partial x_{k_{s+1}}} \left\{[a(x) - a(x^1)]j_\epsilon(x - x^1)\right\} [u(x) - \phi_{,t}(x)]\, dx$$

converges to 0 uniformly for x^1 on D and for fixed $\epsilon > 0$ as $t \longrightarrow \infty$. On the other hand, if $r_{k_2 \ldots k_{s+1}}$ is the limit on $D_{\rho/2}$ of the Cauchy sequence

$$\left\{\frac{\partial^s \phi_{,t}}{\partial x_{k_2} \cdots \partial x_{k_{s+1}}}\right\}$$

in the 0-norm,

$$\|R'_\epsilon(\phi_{,t}) - R'_\eta(\phi_{,t})\|_{o,D_{2\rho}}$$

$$= \left\| \int_D \frac{\partial}{\partial x_{k_1}} \left\{ [a(x) - a(x^1)][j_\epsilon(x - x^1) - j_\eta(x - x^1)] \right\} \frac{\partial^s \phi_t}{\partial x_{k_2} \cdots \partial x_{k_{s+1}}} dx \right\|_{o,D_{2\rho}}$$

$$\leq \left\| M_\epsilon \left(\frac{\partial^s \phi_{,t}}{\partial x_{k_2} \cdots \partial x_{k_{s+1}}} \right) - M_\epsilon \left(r_{k_2 \cdots k_{s+1}} \right) \right\|_{o,D_{2\rho}}$$

$$(3.18) \qquad + \left\| M_\eta \left(\frac{\partial^s \phi_{,t}}{\partial x_{k_2} \cdots \partial x_{k_{s+1}}} \right) - M_\eta \left(r_{k_2 \cdots k_{s+1}} \right) \right\|_{o,D_{2\rho}}$$

$$+ \left\| M_\epsilon \left(r_{k_2 \cdots k_{s+1}} \right) - M_\eta \left(r_{k_2 \cdots k_{s+1}} \right) \right\|_{o,D_{2\rho}}$$

$$\leq K \left\| \frac{\partial^s \phi_{,t}}{\partial x_{k_2} \cdots \partial x_{k_{s+1}}} - r_{k_2 \cdots k_{s+1}} \right\|_{o,D_{2\rho}} + o(\epsilon, \eta)$$

where $o(\epsilon, \eta) \longrightarrow 0$ as $\epsilon, \eta \longrightarrow 0$ independently of t. If we choose ϵ_0 so small that $o(\epsilon, \eta) < \frac{\delta}{5}$ for $\epsilon, \eta < \epsilon_0$ and then t so large that

$$\|R'_\epsilon(u) - R'_\epsilon(\phi_{,t})\|_{o,D_{2\rho}}, \quad \|R'_\eta(u) - R'_\eta(\phi_{,t})\|_{o,D_{2\rho}}$$

and

$$\left\| \frac{\partial^s \phi_{,t}}{\partial x_{k_2} \cdots \partial x_{k_{s+1}}} - r_{k_2 \cdots k_{s+1}} \right\|_{o,D\rho}$$

are less than $\frac{\delta}{5}$, then by the **triangle** inequality for the O-norm, $\|R'_\epsilon(u) - R'_\eta(u)\|_{o,D_{2\rho}}$ is less than δ. Q.E.D.

LEMMA 3. (Rellich). Let $\{u_{,t}\}$ be an infinite sequence from $H_{m,r}(D)$ on the bounded domain D with uniformly bounded m-norm. Then there exists an infinite subsequence $\{u_{,t_1}\}$ of

$\{u_{,t}\}$ which converges in $H_{m-1,r}(D)$.

PROOF. It suffices to prove the lemma for $r = 1$. With each element $u_{,t} \in H_m(D)$ from the sequence, we may associate $\phi_{,t} \in C_c^\infty(D)$ such that $\|u_{,t} - \phi_{,t}\|_m < 1/t$, since $C_c^\infty(D)$ is dense in $H_m(D)$. If we can extract an infinite subsequence convergent in the $(m - 1)$-norm from $\{\phi_{,t}\}$ the corresponding subsequence $\{u_{,t_i}\}$ will satisfy the conditions of the lemma. Further $\|\phi_{,t}\|_m \leq M + 1 = M_1$ where M is the uniform bound for the m-norms of $\{u_{,t}\}$.

If $u \in C_c^\infty(C_0)$, C_0 an n-cube, we have

$$u(x) - u(y)$$

$$= \sum_{i=1}^{n} \int_{x_i}^{y_i} \frac{\partial u}{\partial x_i} (y_1, \ldots, y_{i-1}, t, x_{i+1}, \ldots, x_n)\, dt$$

for $x, y \in C_0$. Thus, if $C_0 = \left\{x \middle| a_i \leq x_i \leq b_i,\ b_i - a_i = d\right\}$, then by Schwarz's inequality,

$$|u(x) - u(y)|^2 \leq 2nd \sum_{i=1}^{n} \int_{a_i}^{b_i} \left|\frac{\partial u}{\partial x_i} (y_1, \ldots, y_{i-1}, t, x_{i+1}, \ldots, x_n)\right|^2 dt$$

Expanding the term on the left and integrating with respect to x and y over C_0 we obtain, again using the Schwarz inequality

$$d^n \int_{C_0} |u(x)|^2\, dx \leq \left|\int_{C_0} u(x)\, dx\right|^2 + nd^{n+2}\|u\|^2_{1,C_o} \tag{3.19}$$

Since D is bounded, it may be enclosed in a n-cube C. For a positive integer r, C may be decomposed into r^n cubes each with side length $1/r$. If $u \in C_c^\infty(D)$, u may be extended to an element of $C_c^\infty(C)$ by setting it equal to 0 outside of D. Applying (3.19) to u in each of the cubes C_j of the decomposition and summing, we obtain

$$\|u\|^2_{o,C} \leq r^n \sum_{j=1}^{r^n} \left[\int_{C_j} u(x)\, dx\right]^2 + nr^{-2}\|u\|^2_{1,C_o} \tag{3.20}$$

By Lemma 1,

$$\frac{\partial^{m-1}\phi_{,t}}{\partial x_{k_1}\ldots\partial x_{k_{m-1}}}$$

have bounded 0-norms on $L_2(C)$ and by the weak compactness of the solid sphere in Hilbert space, we may choose an infinite subsequence $\{\phi_{,t_i}\}$ for which

$$\int_{C_j} \frac{\partial^{m-1}\phi_{,t_i}}{\partial x_{k_1}\ldots\partial x_{k_{m-1}}}\, dx$$

converge as $i \longrightarrow \infty$ for each of the cubes C_j of each of the decompositions corresponding to any integer r. Let $\delta > 0$. Choose r so large that $nM_1^2/r^2 < \delta/2n^m$. Choose i_0 so large that for $i, j \geq i_0$,

$$\int_{C_j} \frac{\partial^{m-1}\{\phi_{,t_i} - \phi_{,t_j}\}}{\partial x_{k_1}\ldots\partial x_{k_{m-1}}} \leq \delta/2r^n n^{m-1}$$

for every set of indices $(k_1, \ldots, k_{m-1})$, and for every n-cube C_j of the r-decomposition of C. By (3.20), it follows that

$$\left\| \phi_{,t_i} - \phi_{,t_j} \right\|_{m-1}^2 \leq \delta$$

Therefore $\{\phi_{,t_i}\}$ is a Cauchy sequence in the $(m - 1)$-norm and our Lemma is proved.

LEMMA 4. (Sobolev). Let D be a domain in E^n, $M = [\frac{n+1}{2}]$. Suppose u is $M + k$ times strongly differentiable in D. Then there exists $v \in C^{k,r}(D)$ such that $u = v$ almost everywhere in D. Given $\delta > 0$, there exist constants $k_1(\delta)$, $k_2(\delta)$ such that for $x \in D_\delta$, $j \leq k$

$$\left| \frac{\partial^j v}{\partial x_{k_1}\ldots\partial x_{k_j}}(x) \right| \leq k_1(\delta)\|v\|_{j+M,D_{\delta/2}} + k_2(\delta)\|v\|_{j,D_{\delta/2}}$$

PROOF. It suffices to consider $r = 1$. If $\Psi \in C^{\infty}(D)$, $\delta > 0$, $x^{0} \in D$, and h_{δ} as defined on p. 15, then

$$\int_{D} h_{\delta} \sum_{k_1 \ldots k_M}^{n} \left\{ d(x, x_0)^{-n} \prod_{i=1}^{M} \left(x_{k_i} - x_{k_i}^{0} \right) \right\} \frac{\partial^{M} \Psi}{\partial x_{k_1} \ldots \partial x_{k_M}} \, dx$$

(3.21)

$$= \sum_{(k)} (-1)^{M-1} \int_{D} \frac{\partial^{M-1}}{\partial x_{k_2} \ldots \partial x_{k_M}} \left\{ h_{\delta} d(x, x_0)^{-n} \prod_{i=1}^{M} \left(x_{k_i} - x_{k_i}^{0} \right) \right\} \frac{\partial \Psi}{\partial x_{k_1}} \, dx$$

Suppose $x^{0} \in D_{2\delta}$. Since

$$\sum_{k_2 \ldots k_M} \frac{\partial^{M-1}}{\partial x_{k_2} \ldots \partial x_{k_M}} \left\{ d(x, x_0)^{-n} \prod_{i=1}^{M} \left(x_{k_i} - x_{k_i}^{0} \right) \right\}$$

$$= (M - 1)! \, d(x, x_0)^{-n} \left(x_{k_1} - x_{k_1}^{0} \right)$$

the integral of (3.21) may be written in the form

$$\int_{D} k(x, x_0) \Psi(x) \, dx$$

(3.22)

$$+ (-1)^{M+1} (M - 1)! \sum_{k_1=1}^{n} \int_{D} h_{\delta} d(x, x_0)^{-n} \left(x_{k_1} - x_{k_1}^{0} \right) \frac{\partial \Psi}{\partial x_{k_1}} \, dx$$

with $k(x, x_0)$ bounded for $x \in D$, $x_0 \in D_{2\delta}$. On the other hand, it follows by excising a small disk about x_0, integrating by parts, and letting the disk shrink to a point, that

$$\int_{D} \sum_{k_1} d(x - x_0)^{-n} \left(x_{k_1} - x_{k_1}^{0} \right) \frac{\partial \Psi}{\partial x_{k_1}} \, dx = \omega_n \Psi(x_0)$$

where ω_n is the area of the (n-1)-dimensional unit sphere. By the Schwarz inequality, since

$$h_\delta \, d(x, x_o)^{-n} \sum_{i=1}^{M} \left(x_{k_i} - x_{k_i}^o \right)$$

is square summable on E^n,

$$(3.23) \qquad |\Psi(x_o)| \leq k_1(\delta) \|\Psi\|_{M, D_{\delta/2}} + k_2(\delta) \|\Psi\|_{o, D_{\delta/2}}$$

By Lemma 2(b), $J_\epsilon u - J_\eta u \longrightarrow 0$ in the s-norm on $D_{\delta/2}$ as $\epsilon, \eta \longrightarrow 0$ for $s \leq M + j$. By the preceding inequality, $J_\epsilon u$ and all its derivatives of order less than j converge uniformly on D_δ to a function v_δ which must lie in $C^j(D_\delta)$ and itself be $M + j$ times strongly differentiable in D_δ. It follows easily that v_δ does not depend on δ and since $J_\epsilon u \longrightarrow u$ in $L_2(D)$, $u = v$ almost everywhere in D. If we define $\|v\|_{s,D_\delta}$ to be the limit as of $\|J_\epsilon v\|_{s,D_\delta}$ as $\epsilon \longrightarrow 0$ the final inequality follows from (3.23).

§4. PROOF OF THEOREM 1

Let

$$K^{(1)} = \frac{K+\overline{K}}{2}, \qquad K^{(2)} = \frac{K-\overline{K}}{2}$$

Then $K = K^{(1)} + K^{(2)}$; and for $\phi, \Psi \in C_c^{\infty, r}(D)$, $K^{(1)}\phi \cdot \Psi = \phi \cdot K^{(1)}\Psi$, $K^{(2)}\phi \cdot \Psi = - \phi \cdot K^{(2)}\Psi$. Therefore $\mathrm{Re}\left\{(-1)^m K \phi \cdot \phi\right\} = (-1)^m K^{(1)} \phi \cdot \phi$.

Since for any s,

$$a(x) \frac{\partial^s u}{\partial x_{k_1} \ldots \partial x_{k_s}}$$

$$= \frac{\partial}{\partial x_{k_s}} \left[a(x) \frac{\partial^{s-1} u}{\partial x_{k_1} \ldots \partial x_{k_{s-1}}} \right] - \frac{\partial a}{\partial x_{k_s}} \frac{\partial^{s-1} u}{\partial x_{k_1} \ldots \partial x_{k_{s-1}}}$$

and since by hypothesis every coefficient $a_{k_1 \ldots k_s; i, j}$ of K belongs to $C^{\gamma(s)}(\overline{D}) \cap C^{s+M}(\overline{D})$, K may be written in the form

(4.1)
$$(Ku)_i = \sum_{\substack{j,(k),(h)\\0\le s,t\le m}} \frac{\partial^s}{\partial x_{k_1}\cdots\partial x_{k_s}}\left(b_{k_1\ldots k_s;h_1\ldots h_t;i,j}(x)\,\frac{\partial^t u_j}{\partial x_{h_1}\cdots\partial x_{h_t}}\right)$$

with $b_{k_1\ldots k_s;h_1\ldots h_t;i,j} \in C^{M+t}(D) \cap C^0(\overline{D})$. It follows that

(4.2)
$$(\overline{K}u)_i = \sum_{\substack{j,(k),(h)\\0\le s,t\le m}} (-1)^{t+s}\frac{\partial^t}{\partial x_{h_1}\cdots\partial x_{h_t}}\left(b_{k_1\ldots k_s;h_1\ldots h_t;j,i}(x)\frac{\partial^s u_j}{\partial x_{k_1}\cdots\partial x_{k_s}}\right)$$

Finally,

$$(K^{(1)}u)_i = \sum_{\substack{j,(k),(h)\\s,t\le m}} \frac{\partial^s}{\partial x_{k_1}\cdots\partial x_{k_s}}\left(c_{k_1\ldots k_s;h_1\ldots h_t;i,j}\,\frac{\partial^t u_j}{\partial x_{h_1}\cdots\partial x_{h_t}}\right)$$

with $c_{k_1\ldots k_s;h_1\ldots h_t} \in C^0(\overline{D})$ while $c_{k_1\ldots k_m;h_1\ldots h_m;i,j} =$ $1/2\left\{a_{k_1\ldots k_m;h_1\ldots h_m;i,j} + a_{k_1\ldots k_m;h_1\ldots h_m;j,i}\right\}$.

For $\phi \in C_c^{\infty,r}(D)$,

$$(-1)^m K^{(1)}\phi\cdot\phi = (\phi,\phi)^{(1)} + \sum_{\substack{i,j,(k),(h)\\s+t<2m}} (-1)^{m+s}\int_D c_{k_1\ldots k_s;h_1\ldots h_t;i,j}(x)\,\frac{\partial^t\phi_j}{\partial x_{h_1}\cdots\partial x_{h_t}}\,\frac{\partial^s\phi_i^*}{\partial x_{k_1}\cdots\partial x_{k_s}}\,dx$$

where, generally,

(4.3)
$$(\phi,\Psi)^{(1)} = \sum_{i,j,(k),(h)} \int_D c_{k_1\ldots k_m;h_1\ldots h_m;i,j}(x)\,\frac{\partial^m\phi_j}{\partial x_{h_1}\cdots\partial x_{h_m}}\,\frac{\partial^m\Psi_i^*}{\partial x_{k_1}\cdots\partial x_{k_m}}\,dx$$

By Lemma 1 and the Schwarz inequality, there exists $k_o \geq 0$ such that the sum of the integrals of the lower order derivatives in $(-1)^m K^{(1)} \phi . \phi$ is less in absolute value than $k_o \|\phi\|_m \|\phi\|_{m-1}$. We shall show first that

$$(\phi, \phi)^{(1)} \geq \rho \|\phi\|_m^2 - k_1 \|\phi\|_m \|\phi\|_{m-1}$$

for suitable $\rho > 0$, $k_1 \geq 0$ and all $\phi \in C_c^{\infty, r}(D)$ and consequently that $(-1)^m K^{(1)} \phi \cdot \phi \geq \rho \|\phi\|_m^2 - (k_o + k_1) \|\phi\|_m \|\phi\|_{m-1}$.

If ξ is a real n-vector, η a complex r-vector $(\eta_j = \psi_j + i\zeta_j)$, then by the uniform strong ellipticity of K on D and the symmetry of $c_{k_1 \ldots k_{2m}; i,j}$ in i and j, for $x^o \in D$,

$$\sum_{i,j,(k)} c_{k_1 \ldots k_{2m}; i,j}(x^o) \xi_{k_1} \ldots \xi_{k_{2m}} \eta_j \eta_i^*$$

$$= \sum_{i,j,(k)} c_{k_1 \ldots k_{2m}; i,j}(x^o) \xi_{k_1} \ldots \xi_{k_{2m}} \left(\psi_i \psi_j + \zeta_i \zeta_j\right)$$

$$(4.4) \qquad = \sum_{i,j,(k)} a_{k_1 \ldots k_{2m}; i,j}(x^o) \xi_{k_1} \ldots \xi_{k_{2m}} \left(\psi_i \psi_j + \zeta_i \zeta_j\right)$$

$$\geq \rho_1 \left[\sum_{(k)} \left(\xi_{k_1} \ldots \xi_{k_m}\right)^2 \right] \sum_{j=1}^{r} \left(\psi_j^2 + \zeta_j^2\right)$$

$$= \rho_1 \sum_{(k)} \left(\xi_{k_1} \ldots \xi_{k_m}\right)^2 \left(\sum_j |\eta_j^2| \right)$$

If $\phi \in C_c^{\infty, r}(D)$, we set the components of ϕ equal to zero outside of D and for each j, consider the Fourier transform of ϕ_j,

$$(4.5) \qquad F_j(\xi) = (2\pi)^{-n/2} \int_{E^n} e^{i \sum_{k=1}^{n} \xi_k x_k} \phi_j(x) \, dx$$

Integrating by parts, we see that the Fourier transform of

$$\frac{\partial^m \phi_j}{\partial x_{k_1} \dots \partial x_{k_m}}$$

is $(-i)^m \xi_{k_1} \dots \xi_{k_m} F_j(\xi)$. Since the Fourier transform is a unitary transformation on $L_2(E^n)$, (cf. [2], pp. 120-121), for $x^o \in D$,

$$\sum_{i,j(k)} \int_D c_{k_1 \dots k_{2m}; i,j}(x^o) \frac{\partial^m \phi_i^*}{\partial x_{k_1} \dots \partial x_{k_m}} \frac{\partial^m \phi_j}{\partial x_{k_{m+1}} \dots \partial x_{k_m}} dx$$

$$= \sum_{i,j(k)} \int_{E^n} c_{k_1 \dots k_{2m}; i,j}(x^o) \xi_{k_1} \dots \xi_{k_{2m}} F_i^*(\xi) F_j(\xi) \, d\xi \tag{4.6}$$

$$\geq \rho_1 \sum_{j,(k)} \int_{E^n} \left(\xi_{k_1} \dots \xi_{k_m}\right)^2 |F_j(\xi)|^2 \, d\xi = \rho_1 \|\phi\|_m^2$$

Since $c_{k_1 \dots k_{2m}; i,j}(x)$ is uniformly continuous on D, there exists $\epsilon > 0$ such that for $d(x, y) < \epsilon$,

$$|c_{k_1 \dots k_{2m}; i,j}(x) - c_{k_1 \dots k_{2m}; i,j}(y)| < \rho_1/2n^{2m} r^2$$

for every combination of indices. By the Schwarz inequality, if $\phi \in C_c^{\infty,r}(D)$ and $\operatorname{diam}(S(\phi)) < \epsilon$ then

$$\left|(\phi, \phi)^{(1)} - \sum_{i,j,(k)} \int_D c_{k_1 \dots k_{2m}; i,j}(x^o) \frac{\partial^m \phi_i^*}{\partial x_{k_1} \dots \partial x_{k_m}} \frac{\partial^m \phi_j}{\partial x_{k_{m+1}} \dots \partial x_{k_{2m}}} dx\right|$$

$$\leq (\rho_{1/2}) \|\phi\|_m^2$$

for any $x^o \in S(\phi)$ and as a consequence

$$(\phi, \phi)^{(1)} \geq \rho_1/2\|\phi\|_m^2$$

Let $\{N_\lambda\}$ be a finite open covering of D by disks of diameter less than ϵ, $\{\ _\lambda^2\}$ a real-valued partition of unity with respect to this covering, i.e., $\mathfrak{x}_\lambda \in C_c^\infty(N_\lambda)$,

$$\sum_{\lambda=1}^{q} \left[\mathfrak{x}_\lambda(x)\right]^2 = 1$$

for $x \in D$. For $\phi \in C_c^{\infty,r}(D)$, we define $(\mathfrak{x}_\lambda \phi)_j = \mathfrak{x}_\lambda \phi_j$. Then $\mathfrak{x}_\lambda \phi \in C_c^{\infty,r}(D \cap N_\lambda)$ and by the preceding paragraph $\left(\mathfrak{x}_\lambda \phi, \mathfrak{x}_\lambda \phi\right)^{(1)} \geq \rho\|\mathfrak{x}_\lambda \phi\|_m^2$, $(\rho = \rho_1/2)$. Since

$$\Psi = \sum_{\lambda=1}^{q} \mathfrak{x}_\lambda^2 \Psi$$

on D,

$$(\phi, \phi)^{(1)} = \sum_{\lambda,i,j,(k)} \int_D c_{k_1 \ldots k_{2m};i,j}(x)\mathfrak{x}_\lambda^2(x) \frac{\partial^m \phi_i^*}{\partial x_{k_1} \ldots \partial x_{k_m}} \frac{\partial^m \phi_j}{\partial x_{k_{m+1}} \ldots \partial x_{k_{2m}}} dx$$

$$= \sum_{\lambda,i,j,(k)} \int_D c_{k_1 \ldots k_{2m};i,j}(x) \frac{\partial^m (\mathfrak{x}_\lambda \phi_i)^*}{\partial x_{k_1} \ldots \partial x_{k_m}} \frac{\partial^m (\mathfrak{x}_\lambda \phi_j)}{\partial x_{k_{m+1}} \ldots \partial x_{k_{2m}}} dx + r_1(\phi)$$

$$(4.7) \qquad \geq \rho \sum_{\lambda,i,j(k)} \int_D \frac{\partial^m (\mathfrak{x}_\lambda \phi_j)}{\partial x_{k_1} \ldots \partial x_{k_m}} \frac{\partial^m (\mathfrak{x}_\lambda \phi_j)^*}{\partial x_{k_1} \ldots \partial x_{k_m}} dx + r_1(\phi)$$

$$= \rho \sum_{\lambda,j,(k)} \int_D \mathfrak{x}_\lambda^2 \left| \frac{\partial^m \phi_j}{\partial x_{k_1} \ldots \partial x_{k_m}} \right|^2 dx + r_1(\phi) + r_2(\phi)$$

$$= \rho\|\phi\|_m^2 + r_1(\phi) + r_2(\phi)$$

where $r_1(\phi)$ and $r_2(\phi)$ involve integrals of products of derivatives of ϕ, one of which at least has order less than m. By Schwarz' inequality and Lemma 1 there exists $k_1 \geq 0$ such that $|r_1(\phi) + r_2(\phi)| \leq k_1\|\phi\|_m\|\phi\|_{m-1}$. It follows as well that there exists $k_2 \geq 0$ for which

$$(4.8)\qquad \begin{aligned} &(a)\quad (-1)^m K^{(1)}\phi\cdot\phi \geq \rho\|\phi\|_m^2 - k_2\|\phi\|_m\|\phi\|_{m-1} \\ &(b)\quad |(-1)^m K^{(1)}\phi\cdot\phi| \leq k_2\|\phi\|_m^2 \end{aligned}$$

Suppose Theorem 1 were not true. Then for every positive integer j, there would exist $\phi^j \in C_c^{\infty,r}(D)$ with $\|\phi^j\|_m = 1$ for which

$$(-1)^m K^{(1)}\phi^j \cdot \phi^j \leq (1/j)\ \|\phi^j\|_m^2 - j\ \|\phi^j\|_0^2$$

By inequality (b), $\|\phi^j\|_0 \longrightarrow 0$ as $j \longrightarrow +\infty$. By Lemma 3, we may extract an infinite subsequence $\{\phi^{j_k}\}$ which converges in the (m-1)-norm to $h \in H_{m-1,r}(D)$. Since this subsequence also converges to 0 in the 0-norm and the mapping of $H_{m-1,r}(D)$ into $L_{2,r}(D)$ is one-one, $h = 0$. Thus $\|\phi^{j_k}\|_{m-1} \longrightarrow 0$ as $k \longrightarrow \infty$. But then

$$1/j_k \geq (-1)^m K^{(1)}\phi^{j_k}\cdot\phi^{j_k} \geq \rho(1 - k_1\ \|\phi^{j_k}\|_{m-1}) \geq \rho/2$$

for k sufficiently large, which is a contradiction. Theorem 1 follows, Q.E.D.

§5. PROOF OF THEOREM 2

Proof of Theorem 2. By hypothesis u is m-times strongly differentiable in D. We shall assume that the coefficients $a_{k_1\ldots k_s;i,j}$ of K lie in $C^{M+s+t_o}(D)$ for some fixed $t_o \geq 0$ while γ is $M+t_o$ times strongly differentiable in D, and shall show that u is $M+2m+t_o$ times strongly differentiable in D. Suppose that we have shown for a given integer $t \leq M + m + t_o$, that u is $m+t-1$ times strongly differentiable in D. Then if $\delta > 0$, $J_\epsilon(u) - J_\eta(u) \longrightarrow 0$ in the (m+t-1)-norm on $D_{\delta/2}$ as $\epsilon, \eta \longrightarrow 0$. Let h_δ be the function from $C_c^\infty(D_{\delta/2})$ defined in §3 such that $0 \leq h(x) \leq 1$, $h_\delta(x) = 1$ on D_δ. We define the auxiliary system K^+ by,

$$(K^+u)_i = \sum_{j,(k)} (-1)^s a_{k_1\ldots k_s;j,i}(x)\ \frac{\partial^s u_j}{\partial x_{k_1}\ldots\partial x_{k_s}}$$

It follows from the definition that

$$(\overline{K^+ u})_i = \sum_{j,(k)} (-1)^s e_{k_1 \ldots k_s; j, i}(x) \frac{\partial^s u_j}{\partial x_{k_1} \ldots \partial x_{k_s}}$$

where $e_{k_1 \ldots k_s; i, j}$ is the corresponding coefficient in $\overline{K}$ and lies in $C^{M+2m+t}{}_0(D)$. Let

$$(\Delta^t u)_i = \left(\sum_{k=1}^{n} \frac{\partial^2}{\partial x_k{}^2} \right)^t u_i$$

Then $K^+ \Delta^t$ is uniformly strongly elliptic on every compact subset of D and by Theorem 1 there exist $\rho_t > 0$, $k_t \geq 0$ such that for $\phi \in C_c^{\infty, r}(D_{\delta/2})$

(5.1) $$\operatorname{Re}\left\{(-1)^{m+t} K^+ \Delta^t \phi \cdot \phi\right\} \geq \rho_t \|\phi\|^2_{m+t} - k_t \|\phi\|^2_0$$

Since for $\epsilon, \eta > 0$, $h_\delta(J_\epsilon u - J_\eta u)$ lies in $C_c^{\infty, r}(D_{\delta/2})$ we have

$$\|h_\delta(J_\epsilon u - J_\eta u)\|^2_{m+t} \leq (1/\rho_t) \left| (K^+ \Delta^t h_\delta(J_\epsilon u - J_\eta u) \cdot h_\delta(J_\epsilon u - J_\eta u) \right|$$

(5.2)
$$+ (k_t/\rho_t) \|J_\epsilon u - J_\eta u\|^2_{0, D_{\delta/2}}$$

If we integrate by parts, $(K^+ \Delta^t h_\delta(J_\epsilon u - J_\eta u) \cdot h_\delta(J_\epsilon u - J_\eta u)) = (\Delta^t h_\delta(J_\epsilon u - J_\eta u) \cdot \overline{K^+} h_\delta(J_\epsilon u - J_\eta u)) = (\Delta^t h_\delta(J_\epsilon u - J_\eta u) \cdot h_\delta K^+(J_\epsilon u - J_\eta u)) + r_3(J_\epsilon u - J_\eta u)$ where r_3, being composed of terms in which h has been differentiated and in which the sum of the orders of differentiation of u is less than $2m + 2t$, satisfies the inequality

$$|r_3(J_\epsilon u - J_\eta u)|$$

$$\leq k_4 \|J_\epsilon u - J_\eta u\|^1_{m+t-1, D_{\delta/2}} \left\{ \|h_\delta(J_\epsilon u - J_\eta u)\|_{m+t} + \|J_\epsilon u - J_\eta u\|^1_{m+t-1, D_{\delta/2}} \right\}$$

(We have set $\|w\|^1_s = \sum_{0 \leq t \leq s} \|w\|_t$.)

Thus

$$r_3 < r_3'(\epsilon, \eta) \, \|h_\delta(J_\epsilon u - J_\eta u)\|_{m+t} + r_3''(\epsilon, \eta)$$

where $r_3, r_3' \longrightarrow 0$ as $\epsilon, \eta \longrightarrow 0$ by the inductive hypothesis of Lemma 2(b).

On the other hand, for $x^1 \in D_{\delta/2}$,

$$(\overline{K^+ J_\epsilon u})_i(x^1) = \sum_{j,(k)} (-1)^s e_{k_1 \ldots k_s; j, i}(x^1) \frac{\partial^s}{\partial x^1_{k_1} \ldots \partial x^1_{k_s}} \int_D j_\epsilon(x - x^1) u_j(x)\, dx$$

$$= \int_D \sum_{j,(k)} e_{k_1 \ldots k_s; j, i}(x) \frac{\partial^s (j_\epsilon(x - x^1))}{\partial x_{k_1} \ldots \partial x_{k_s}} u_j(x)\, dx + \sum_{j,(k)} R_{\epsilon, k_1 \ldots k_s; j, i}(x^1)$$

where

$$R_{\epsilon, k_1 \ldots k_s; i, j}(x^1) = \int_D \left[e_{k_1 \ldots k_s; j, i}(x^1) - e_{k_1 \ldots k_s; j, i}(x) \right] \frac{\partial^s (j_\epsilon(x - x^1))}{\partial x_{k_1} \ldots \partial x_{k_s}} u_j(x)\, dx$$

If we consider $(\Delta^t h_\delta(J_\epsilon u - J_\eta u) \cdot h_\delta \overline{K^+}(J_\epsilon u - J_\eta u))$, the contribution of the terms of the first sum of the above decomposition will be, (setting $\alpha_i(x^1) = \Delta^t h_\delta(J_\epsilon u - J_\eta u)_i(x^1)$)

$$\sum_{i, j(k)} \int_D \int_D e_{k_1 \ldots k_s; j, i}(x) \frac{\partial^s (\alpha_i(x^1) j_\epsilon(x - x^1) - j_\eta(x - x^1))}{\partial x_{k_1} \ldots \partial x_{k_s}} u_j^*(x)\, dx\, dx^1$$

$$= (\Delta^t h_\delta(J_\epsilon u - J_\eta u) \cdot h_\delta(J_\epsilon \gamma - J_\eta \gamma))$$

provided $\epsilon, \eta < \delta/2$. This follows from the fact that u is assumed to be a weak solution of the system $Ku = \gamma$ and that

$$\sum_{i,(k)} e_{k_1 \ldots k_s; j, i}(x) \frac{\partial^s (\alpha_i(x^1)[j_\epsilon(x - x^1) - j_\eta(x - x^1)]}{\partial x_{k_1} \ldots \partial x_{k_s}}$$

$$= \overline{K}(\alpha_i[j_\epsilon(x - x^1) - j_\eta(x - x^1)])$$

If $t \leq m$, the sum of these terms is less in absolute value than $k_5\|h_\delta(J_\epsilon u - J_\eta u)\|_{m+t}\ \|h_\delta(J_\epsilon \gamma - J_\eta \gamma)\|_0$. If $t > m$, the difference may be shown by $t-m$ integrations by parts to be less in absolute value than $k_5\|h_\delta(J_\epsilon u - J_\eta u)\|_{m+t}\ \|J_\epsilon \gamma - J_\eta \gamma\|_{t-m,D_{\delta/2}}$. By the assumption on the differentiability of γ and Lemma 2(b), in both cases the factor multiplying $\|h_\delta(J_\epsilon u - J_\eta u)\|_{m+t}$ goes to zero as $\epsilon, \eta \longrightarrow 0$.

Let $I_{i,j,(k)} = (\Delta^t h_\delta(J_\epsilon u_i - J_\eta u_i) \cdot (R_{\epsilon,k_1\ldots k_s;j,i} - R_{\eta,k_1\ldots k_s;j,i}))$

For $t > m$, integrating by parts $t-m$ times, I is the sum of terms of the form,

$$\frac{\partial^{t+m}}{\partial x^1_{p_1}\ldots\partial x^1_{p_{t+m}}}\left\{h_\delta(J_\epsilon u - J_\eta u)\right\}\frac{\partial^{t-m}}{\partial x^1_{q_1}\ldots\partial x^1_{q_{t-m}}}\int_D [d(x)-d(x^1)]\ \frac{\partial^s[j_\epsilon(x-x^1)-j_\eta(x-x_1)]}{\partial x_{k_1}\ldots\partial x_{k_s}}u^*_j(x)\ dx$$

with $d \in C^{M+s+t}(D)$. Carrying the differentiations under the integral sign, we see that such terms are bounded in absolute value by

$$k_6\ \|h_\delta(J_\epsilon u - J_\eta u)\|_{m+t}\left\{\|J_\epsilon u - J_\eta u\|_{m+t-1,D_{\delta/2}} + \left\|\int_D [d(x) - d(x^1)]\right.\right.$$

$$\left.\left.\frac{\partial^{t-m+s}[j_\epsilon(x-x^1)-j_\eta(x-x^1)]}{\partial x_{k_1}\ldots\partial x_{k_s}\partial x_{q_1}\ldots\partial x_{q_{t-m}}}\ u^*_j(x)\ dx\right\|_{0,D_{\delta/2}}\right.$$

The sum of such terms in absolute value is less than $\|h_\delta(J_\epsilon u - J_\eta u)\|_{m+t}r_5(\epsilon, \eta)$ where $r_5(\epsilon, \eta)$ approaches zero as $\epsilon, \eta \longrightarrow 0$ by the inductive hypothesis and Lemma 2(b) and (c). For $t \leq m$, $s \leq m + t$, I may be immediately estimated by the Schwarz inequality as less in absolute value than

$$k_7\|h_\delta(J_\epsilon u - J_\eta u)\|_{m+t}\left\|\int_D [d(x) - d(x^1)]\ \frac{\partial^s[j_\epsilon(x-x^1)-j_\eta(x-x^1)]}{\partial x_{k_1}\ldots\partial x_{k_s}}\ dx\ \right\|_{0,D_{\delta/2}}$$

$\leq \|h_\delta(J_\epsilon u - J_\eta u)\|_{m+t}r_6(\epsilon, \eta)$ where $r_6 \longrightarrow 0$ as $\epsilon, \eta \longrightarrow 0$ according to the inductive hypothesis and Lemma 2(c). If $t \leq m$, $s > m + t$, we may pull out $s - m - t$ derivatives from inside the integral and (using as before the fact that

$$\frac{\partial}{\partial x_i} j_\epsilon(x - x^1) = -\frac{\partial}{\partial x_i^1} j_\epsilon(x - x^1))$$

I may be written as the sum of terms of the form

$$\frac{\partial^{m+t}}{\partial x_{p_1} \ldots \partial x_{p_{m+t}}} h_\delta(J_\epsilon u_i - J_\eta u_i) \cdot f(x^1) \frac{\partial^{s_1}}{\partial x_{q_1} \ldots \partial x_{q_{s_1}}} (J_\epsilon u_j - J_\eta u_j)$$

with $f \in C^0(D)$ and $s_1 \leq m + t$ as well as terms of the form

$$\frac{\partial^{m+t} h_\delta(J_\epsilon u_i - J_\eta u_i)}{\partial x^1_{p_1} \ldots \partial x^1_{p_{m+t}}} \cdot \int_D [d(x) - d(x^1)] \frac{\partial^{m+t}[j_\epsilon(x-x^1) - j_\eta(x-x^1)]}{\partial x_{q_1} \ldots \partial x_{q_{m+t}}} \quad dx$$

with $d \in C^{m+t}(D)$. Again as before, I is less than

$$r_7(\epsilon, \eta) \| h_\delta(J_\epsilon u - J_\eta u) \|_{m+t}$$

with $r_7 \longrightarrow 0$ as $\epsilon, \eta \longrightarrow 0$.

If we sum all the terms of the right hand side of the inequality,

$$\| h_\delta(J_\epsilon u - J_\eta u) \|^2_{m+t} \leq r_8(\epsilon, \eta) \| h_\delta(J_\epsilon u - J_\eta u) \|_{m+t} + r_8(\epsilon, \eta)$$

with $r_8 \longrightarrow 0$ as $\epsilon, \eta \longrightarrow 0$. It follows easily that

$$\| h_\delta(J_\epsilon u - J_\eta u) \|_{m+t} \longrightarrow 0$$

as $\epsilon, \eta \longrightarrow 0$, and therefore $\| J_\epsilon u - J_\eta u \|_{m+t, D_\delta}$, which is even smaller, converges to zero as $\epsilon, \eta \longrightarrow 0$. Thus u is m+t times strongly differentiable in D since δ is arbitrary. It follows by a repetition of this argument that u is $(M+2m+t_0)$-times strongly differentiable in D and hence by Lemma 2 equals v almost everywhere in D for some $v \in C^{2m+t_0}(D)$. Integrating the equation $v \cdot \overline{K}\phi = \gamma \cdot \phi$ by parts, $(Kv - \gamma) \cdot \phi = 0$ for all $\phi \in C_c^{\infty, r}(D)$ and hence $Kv = \gamma$ in D. Q.E.D.

§6. THE DIRICHLET PROBLEM

Proof of Theorem 3. By Theorem 1, for all $\phi \in C_c^{\infty,r}(D)$, if k_0 is chosen sufficiently large, $(-1)^m K^{(1)} \phi \cdot \phi + k_0 \phi \cdot \phi \geq \rho \|\phi\|_m^2$. If for $\phi, \Psi \in C_c^{\infty,r}(D)$, we define the inner product (ϕ, Ψ) by $(\phi, \Psi) = (-1)^m K^{(1)} \phi \cdot \Psi + k_0 \phi \cdot \Psi$, it follows from the Schwarz inequality and Lemma 1 that there exists $c_0 > 0$ for which

$$c_0 \|\phi\|_m^2 \leq (\phi, \phi) \leq 1/c_0 \|\phi\|_m^2$$

As a result, (ϕ, Ψ) may be extended by continuity to $H_{m,r}(D)$ and yields a Hilbert space structure on that set which has a norm equivalent to the m-norm. We designate this Hilbert space by $H(D)$.

For $u \in H(D)$,

$$|u \cdot \phi| \leq c_1 \|u\|_0 \|\phi\|_0 \leq c_2 \, d^m \|u\|_0 \|\phi\|_m \leq c_3 \, d^{2m} \|u\|_m \|\phi\|_m$$

by the Schwarz inequality and Lemma 1 for all $\phi \in C_c^{\infty,r}(D)$. Holding u fixed, $(u \cdot \phi)$ is a bounded, conjugate-linear complex functional on the dense subset $C_c^{\infty,r}(D)$ of $H(D)$ and, since the latter is a Hilbert space, by the Frechet-Riesz Theorem, there is a unique element w of $H(D)$ such that $(w, \phi) = (u \cdot \phi)$ for $\phi \in C_c^{\infty,r}(D)$. If we set

$$Uu = w, \quad |(Uu, \phi)| \leq c_2 \, d^m \|u\|_0 \|\phi\|_m$$

and as a consequence,

$$\|Uu\| \leq c_2' \, d^m \|u\|_0 \leq c_3 \, d^m \|u\|_m$$

In addition,

$$(U\phi, \Psi) = \phi \cdot \Psi = (\Psi \cdot \phi)^* = (U\Psi, \phi)^* = (\phi, U\Psi)$$

for $\phi, \Psi \in C_c^{\infty,r}(D)$. It follows easily that U is a bounded Hermitian linear transformation of $H(D)$, while by Lemma 3, since every bounded subset in the m-norm is precompact in the 0-norm, U is completely continuous. Applying the same argument, since

$$|K^{(2)} \phi \cdot \Psi| \leq c_4 \|\phi\|_m \|\Psi\|_m$$

there exists a bounded skew-Hermitian linear transformation W of $H(D)$ such that $(W\phi, \Psi) = (K^{(2)}\phi \cdot \Psi)$ for $\phi, \Psi \in C_c^{\infty,r}(D)$. Similarly if $g \in C^{m,r}(D) \cap L_{2,r}(D)$, g having square-summable m-th derivatives on D, there exists $w^1 \in H(D)$ such that $(w^1, \phi) = (g, \phi)$ for $\phi \in C_c^{\infty,r}(D)$.

If $u \in H(D) \cap C^{2m,r}(D)$, let $\{\phi^j\}$ be a sequence from $C_c^{\infty,r}(D)$ converging to u in $H(D)$. If $\phi \in C_c^{\infty,r}(D)$,

$$(u, \phi) = \lim_j (\phi^j, \phi) = \lim_j \left\{(-1)^m(K^{(1)}\phi^j \cdot \phi) + k_0(\phi^j \cdot \phi)\right\}$$

$$= \lim_j (-1)^m(\phi^j \cdot K^{(1)}\phi) + k_0(\phi^j \cdot \phi)$$

$$=(-1)^m(u \cdot K^{(1)}\phi) + k_0(u \cdot \phi) = ((-1)^m K^{(1)}u + k_0 u \cdot \phi)$$

Similarly $(Wu, \phi) = (K^{(2)}u \cdot \phi)$ while $(Uu, \phi) = u \cdot \phi$. By a part of the previous argument, we may deduce that, if u merely lies in $H(D)$, then $(u, \phi) = (u \cdot (-1)^m K^{(1)}\phi + k_0\phi)$ for $\phi \in C_c^{\infty,r}(D)$ while $(Wu, \phi) = -(u \cdot K^{(2)}\phi)$.

It follows now that $u_1 = w^1 - g$, satisfies $(u, \phi) = 0$ for $\phi \in C_c^{\infty,r}(D)$, where w^1 and g are related as above; i.e., u is a weak solution of the equation $[(-1)^m K^{(1)} + k_0]u = 0$. Since u_1 is m-times strongly differentiable in D, after a change on a set of measure zero in D, u_1 may be taken in $C^{2m,r}(D)$ by Theorem 2 with $[(-1)^m K^{(1)} + k_0]u = 0$. Furthermore, it follows from the proof of Theorem 2 that u_1 is $2m+M$ times strongly differentiable in D.

Since u_1 has square-integrable m-th derivatives, $|K^{(2)}u_1 \cdot \phi| \leq c_5\|\phi\|_m$ for $\phi \in C_c^{\infty,r}(D)$, and $(\xi \cdot \phi) = (\zeta, \phi)$ for a suitable $\zeta \in H(D)$ with $\xi = \gamma - K^{(2)}u_1 - (-1)^m k_0 u_1$. If we set $v = u - u_1$, the Dirichlet problem for $Ku = \gamma$ with boundary function g may be reduced to the Dirichlet problem with zero boundary values for $Kv = \xi$; ξ is M-times strongly differentiable in D. By Theorem 2, any function v in $H(D)$ lies (after a change on a set of zero measure) in $C^{2m,r}(D)$ and v is a solution of the equation $Kv = \xi$ if and only if $(v \cdot \overline{K}\phi) = (\xi \cdot \phi)$ for all $\phi \in C_c^{\infty,r}(D)$; this is equivalent to $(\zeta, \phi) = (-1)^m(v \cdot [(-1)^m K^{(1)}\phi + k_0\phi]) - (v \cdot K^{(2)}\phi) - (-1)^m k_0(v \cdot \phi) = ([(-1)^m v + Wv - (-1)^m k_0 Uv], \phi)$. The latter equation holds for all $\phi \in C_c^{\infty,r}(D)$ if and only if $v + (-1)^m Wv - k_0 Uv = (-1)^m \zeta$.

For $v \in C_c^{\infty,r}(D)$, $\operatorname{Re}\{(I + (-1)^m W)v, v\} \geq c_0\|v\|_m^2$, and by continuity the inequality is true for $v \in H(D)$. It follows that only the zero element of $H(D)$ can lie in the orthogonal complement of the range of $I + (-1)^m W$ so that the transformation $I + (-1)^m W$ in $H(D)$ is onto. Obviously it is one to one. If $R = (I + (-1)^m W)^{-1}$, R is a bounded linear transformation with $\|R\| \leq 1/c_0$. (R is equal to the identity transformation if $K^{(2)} = 0$.) Since RU is completely continuous if U is completely continuous, it follows by an application of the Riesz-Schauder theory of linear completely continuous operators (Cf. [1]), that the equation $v - k_0 RU_v = (-1)^m R\zeta$ has a solution in $H(D)$ if there exist no solutions $u \neq 0$ of $u - k_0 RUu = 0$. By the above argument, we see that any solution u of the homogeneous equation is, after a change on a set of measure zero, a null-solution of the equation $Ku = 0$.

We note that $R^* = (I - (-1)^m W)^{-1}$, from which it follows that an element v_0 of $H(D)$ lies in the null space of

$$(I - k_0 RU)^* = (I - k_0 UR^*) = (I - k_0 U(I - (-1)^m W)^{-1})$$

if and only if $w = (I - (-1)^m W)^{-1} v_0$ satisfies the functional equation $w - (-1)^m Ww - k_0 Uw = 0$. Using Theorem 2, we may easily show as above that, after a change on a set of measure zero in D, w will be a null solution of the equation $\overline{K}w = 0$. It follows, since R is a non-singular operator, that the linear dimensions of the families of null solutions of K and $\overline{K}$ on D (which are both finite integers) must be the same. Furthermore, the equation $Ku = \gamma$ will have a solution u in $H(D)$ if and only if letting ζ be that element of $H(D)$ such that $(\zeta, \phi) = (\gamma \cdot \phi)$ for all $\phi \in C_c^{\infty, r}(D)$, $(R\zeta, R^{*-1}w) = 0$ for all null solutions w of $\overline{K}w = 0$. But $(R\zeta, R^{*-1}w) = (\zeta, w) = (\gamma \cdot w)$.

We note, finally, that $\|U\| \leq d^{2m} c_2$ while $\|RU\| \leq d^{2m} c_2 / c_0$ where d is the diameter of the domain D. If d is sufficiently small for a subdomain D_1 of D, since c_2 and c_0 can be chosen independently of the sub-domain, $(I - k_0 RU)$ will have a trivial null-space and the Dirichlet problem for K on D_1 will always have a solution.

Proof of Theorem 4. By the argument of the proof of Theorem 3, the discussion of the eigenvalues and eigenfunctions of K on D, having first been transformed into the discussion of the eigenvalues and eigenfunctions of $(K + (-1)^m k_0 I)$ on D, may be reduced to the study of the characteristic values and characteristic functions of the operator RU on $H(D)$ since $(K + (-1)^m k_0)u = \lambda u$ for $u \in H(D)$ if and only if $(-1)^m \{(-1)^m u + Wu\} = \lambda Uu$, i.e., if and only if $u = (-1)^m \lambda RUu$. Thus (a) follows from the Riesz theory while (b) follows from the spectral theorem for completely continuous Hermitian operators on a Hilbert space if we observe that U is Hermitian and completely continuous while for K self-adjoint, $K^{(2)} = 0$ and $R = I$.

§7. PROOF OF THEOREM 5

Let u be the solution of the Dirichlet problem for $\overline{K}u = \xi$ on D with zero boundary function. Let $\delta > 0$ and let h_δ be the function applied in the proof of Theorem 2. Then

$$\|h_\delta u\|^2_{m+t,D_\delta} \leq c_6 |(h_\delta u \cdot K\Delta^t h_\delta u)| + c_7 \|h_\delta u\|^2_0$$

$$\leq c_8 (h_\delta \xi \cdot \Delta^t h_\delta u) + c_9 \|h_\delta u\|_{m+t} \|u\|'_{m+t-1,D_{\delta/2}}$$

It follows that there exists $c_{10} > 0$ (depending on δ) such that

$$\|u\|_{m+t,D_\delta} \leq \|h_\delta u\|_{m+t} \leq c_{10} \left\{ \|u\|'_{m+t-1,D_{\delta/2}} + \|\xi\|_{\gamma(t),D_{\delta/2}} \right\}$$

If we recall that j-norms of u on D for $j \leq m$ are bounded by a constant multiple of the 0-norm of ξ on D and apply the last inequality successively for t from 1 up to $M - m$, we see that

$$\|u\|'_{M,D_\delta} \leq c_{11} \|\xi\|_{\gamma(M-m)}$$

Since δ is an arbitrary positive number, it follows from Lemma 4 that for each $x \in D$, there is a constant $c_{11}(x)$ for which

$$|u_i(\xi)(x)| \leq c_{11}(x) \|\xi\|_{\gamma(M-m)}$$

Let $s_0 = \gamma(M - m)$. Since $u_i(\xi)(x)$ for fixed x is a bounded linear functional on a dense subset of $H_{s_0,r}(D)$, there exists $h_{x,i} \in H_{s_0,r}(D)$ for which $u_i^*(\xi)(x) = (h_{x,i}, \xi)_{s_0}$ for all $\xi \in H_{s_0,r}(D)$. In particular if $\phi \in C_c^{\infty,r}(D - \{x\})$, $u(\overline{K}\phi)(x) = 0 = (h_{x,i}, \overline{K}\phi)_{s_0} = (h_{x,i} \cdot \Delta^{s_0}\overline{K}\phi)$. The latter equation signifies, however, that on $D - \{x\}$, $h_{x,i}(z)$ is a weak solution of $K\Delta^{s_0}u = 0$ and, after a change on a set of measure zero, lies in $C^{2m+2s_0,r}(D - \{x\})$. If $g_{x,i}(z) = \Delta_z^{s_0}h_{x,i}(z)$, $(g_{x,i} \cdot \overline{K}\phi) = 0$ for $\phi \in C_c^{\infty,r}(D - \{x\})$. Furthermore, $u_i(\phi)(x) = (g_{x,i} \cdot \phi)$ for $\phi \in C_c^{\infty,r}(D - \{x\})$. Since $s_0 < M$, it is easily shown that $C_c^{\infty,r}(D) \subset H_{s_0,r}(D - \{x\})$. Since $u_i(\phi)(x)$ is continuous in the s_0-norm of ϕ, it follows that $u_i(\phi)(x) = (g_{x,i} \cdot \phi)$ for $\in C_c^{\infty,r}(D)$. In particular $\{g_i(x, z)\}$ is a fundamental solution matrix for K on D. The differentiability in x follows from the fact (which is easily established) that $\{g_{ji}(z, x)\}$ is the Green's matrix for $\overline{K}$ on D. Q.E.D.

BIBLIOGRAPHY

[1] BANACH, S., "Théorie des opérations linéaires," Z subwencji funduszu kultury narodowej. Monografje matematyczne I, Warsaw (1932), 145-164.

[2] BOCHNER, S., and CHANDRASEKHARAN, K., "Fourier transforms," (Annals of Mathematics Study No. 19), Princeton (1949).

[3] BROWDER, F. E., "The Dirichlet problem for linear elliptic equations of arbitrary even order with variable coefficients," Proceedings of the National Academy of Sciences 38 (1952), 230-235.

[4] BROWDER, F. E., "The Dirichlet and vibration problems for linear elliptic differential equations of arbitrary order," Proceedings of the National Academy of Sciences 38 (1952), 741-747.

[5] BROWDER, F. E., "Assumption of boundary values and the Green's function in the Dirichlet problem for the general linear elliptic equation," to appear in the Proceedings of the National Academy of Sciences.

[6] BROWDER, F. E., "Linear parabolic differential equations of arbitrary order; general boundary value problems for elliptic equations," to appear in the Proceedings of the National Academy of Sciences.

[7] BROWDER, F. E., "On the eigenfunctions and eigenvalues of the general linear elliptic differential operators," to appear in the Proceedings of the National Academy of Sciences.

[8] BROWDER, F. E., "Linear elliptic systems of differential equations," to appear.

[9] COURANT, R., and HILBERT, D., Methoden der mathematischen Physik, 2 Springer, Berlin (1937).

[10] DE RHAM, G., and KODAIRA, K., Harmonic integrals, Notes on Lectures at the Institute for Advanced Study, Princeton (1950).

[11] FRIEDRICHS, K. O., "Randwert- und Eigenwertprobleme aus der Theorie der elastischen Platten," Mathematische Annalen 98 (1927), 206-247.

[12] FRIEDRICHS, K. O., "On differential operators in Hilbert spaces," American Journal of Mathematics 61 (1939), 523-544.

[13] FRIEDRICHS, K. O., "The identity of weak and strong extensions of differential operators," Transactions of the American Mathematical Society 55 (1944), 132-151.

[14] FRIEDRICHS, K. O., "On the boundary-value problems of the theory of elasticity and Korn's inequality," Annals of Mathematics 48 (1947), 441-471.

[15] FRIEDRICHS, K. O., "On the differentiability of solutions of linear elliptic differential equations," Communications on Pure and Applied Mathematics, Vol. VI (1953), 299-325.

[16] GÅRDING, L., "Le problème de Dirichlet pour les équations aux dérivées partielles elliptiques homogènes à coefficients constants," Comptes Rendus de l'Académie des Sciences, Paris 230 (1950), 1030-1032.

[17] GÅRDING, L., "Dirichlet's problem and the vibration problem for linear elliptic partial differential equations with constant coefficients," Proceedings of the Symposium on Spectral Theory and Differential Problems, Stillwater (1951), 291-295.

BIBLIOGRAPHY

[18] GÅRDING, L., "Le problème de Dirichlet pour les équations aux dérivées partielles elliptiques linéaires dans les domaines bornés," Comptes Rendue de l'Académie des Sciences, 233 (1951), 1554-1556.

[19] JOHN, F., "General properties of solutions of linear elliptic partial differential equations," Proceedings of the Symposium on Spectral Theory and Differential Problems, Stillwater (1951), 113-175.

[20] JOHN, F., "Elementary expressions for the derivatives of weak solutions of elliptic differential equations," Abstract 632, Bulletin of the American Mathematical Society 58 (1952), 640.

[21] KONDRASHOV, V. I., "On some properties of functions from the spaces L_p." Doklady Akademiia Nauk 48 (1945), 563-566. (Russian).

[22] MORREY, C. B., Multiple integral problems in the calculus of variations, University of California Press, Berkeley (1943).

[23] PETROVSKY, I. G., "Sur l'analyticité des solutions des systèmes d'équations differentielles," Matematicheskii Sbornik 5 (1939), 3-70.

[24] SCHWARTZ, L., Théorie des distributions, 1 and 2, Hermann, Paris (1950).

[25] SOBOLEV, S. L., "On a boundary value problem for the polyharmonic equations," Matematicheskii Sbornik 2 (1937), 467-500. (Russian).

[26] SOBOLEV, S. L., "On a theorem of functional analysis," Matematicheskii Sbornik 4 (1938), 471-497. (Russian).

[27] VISIK, M. I., "The method of orthogonal and direct decomposition in the theory of elliptic differential equations," Matematicheskii Sbornik 25 (1949), 189-234. (Russian).

[28] VISIK, M. I., "On strongly elliptic systems of differential equations," Doklady Akademiia Nauk 74 (1950), 881-884. (Russian).

[29] VISIK, M. I., "On strongly elliptic systems of differential equations," Matematicheskii Sbornik 29 (1951), 615-676. (Russian).

[30] VITZADZE, A. V., "On the uniqueness of the Dirichlet problem for elliptic partial differential equations," Uspekhi Matematicheskikh Nauk 3 (1948), 211-212. (Russian).

[31] WEYL, H., "The method of orthogonal projection in potential theory," Duke Mathematical Journal 7 (1940), 414-444.

III. DERIVATIVES OF SOLUTIONS OF LINEAR ELLIPTIC PARTIAL DIFFERENTIAL EQUATIONS[1,2]

F. John

The differential equations considered here can be written in the form

$$L[u] = P(D_1, \dots, D_n)u = f(x_1, \dots, x_n) \tag{1}$$

Here D_i stands for the differential operator $\partial/\partial x_i$; P is a polynomial of degree $2m$ in the D_i with coefficients which are functions of the independent variables $x_1, \dots, x_n$ and f is a given function. The polynomial P can be thought of as a sum of homogeneous polynomials with degrees varying from $2m$ down to 0. Here the polynomial consisting of the terms of the highest degree $2m$ in P is to be called the "principal part" of P, and will be denoted by $Q(D_1, \dots, D_n)$. For fixed $x_1, \dots, x_n$ the expression $Q(\xi_1, \dots, \xi_n)$ considered as a function of the variables ξ_i is the <u>characteristic form</u> of the differential equation at the point $(x_1, \dots, x_n) = x$. The equation is <u>elliptic</u>, if the form Q is definite for all x in question. In this case the order of the equation is necessarily even, and hence m is an integer.

It appears natural to consider the differential equation (1) only for functions for which the equation is defined immediately, say for the class C_{2m} consisting of the functions for which all derivatives of order $\leq 2m$ exist and are continuous. One of the most remarkable facts about <u>elliptic</u> equations is that a solution u of class C_{2m} of such an equation automatically possesses higher derivatives, if the function f and the coefficients of the equation are sufficiently regular. As an extreme instance, one has the theorem that in case f and the coefficients of the equation are analytic, all solutions are analytic in their domain of

[1] This work was performed under the sponsorship of the Office of Naval Research.

[2] A more complete exposition of the contents of this paper (with closer attention to the exact regularity assumptions made) is given in [3]. The method of proof used is a modification of methods previously applied by the author to establish analyticity of solutions of linear analytic equations and existence of higher derivatives for solutions of general non-linear elliptic equations. (See [4], pp. 234 et seq., and [5] pp. 162-175.)

definition. This is suggested already by the absence of any real characteristic surfaces for an elliptic equation, which by the theorem of Cauchy-Kowalewski excludes the possibility of "piece-wise analytic" solutions of class C_{2m}. At the other extreme we have theorems of the type proved by F. Browder [1] and K. O. Friedrichs [2] which state that, under rather mild regularity assumptions on the coefficients, a measurable function u which satisfies (1) in a "generalized sense" will automatically possess a certain number of derivatives and be a "strict" solution of (1).[3]

Regularity properties of u are most easily established if a suitable expression for the solution u of (1) is known. Such expressions can be obtained from Green's formula, if a fundamental solution of the adjoint differential equation is known. In this case u at a point of a domain D can be expressed in terms of f in D and of the Cauchy data of u on the boundary of D by means of integrals which involve the fundamental solution in their kernel. Given sufficient regularity conditions of the fundamental solution, corresponding regularity properties of u can be established. (See [6].) Fundamental solutions for equations with sufficiently regular coefficients can be constructed,e.g. by the parametrix method. Considerable difficulties arise naturally, if the solutions considered are of the generalized or "weak" types to which Green's identity cannot be applied so easily.

It is possible to obtain similar results in other ways not using actual special solutions of the given differential equation or its adjoint, but taking advantage more directly of the elliptic character of the solution and of its general form. The equation yields directly a variety of integral identities and inequalities. These can be arranged so as to give information on higher derivatives, in terms of lower ones. Here again the use of such inequalities or identities becomes more involved, if existence of these higher derivatives is not know a priori. This is the direction of proof followed in the work of K. O. Friedrichs and in the present paper. Friedrichs works essentially with inequalities and with function spaces associated naturally with equations of the type (1). In the present paper an attempt is made to get some results of this type (which are,however, not as general as those obtained by the other authors) in as elementary a fashion as possible, by using only integral identities instead of estimates, and by considering only continuous solutions.

The results given here are based on the derivation of an integral expression for the derivatives of a solution u in terms of u itself. This expression is "elementary" in that it can be obtained by "elementary operations" on the coefficients of the equation. Its derivation yields at

---

[3] Similar theorems for non-linear equations are more difficult to establish. They can be derived from more refined results for the linear case, like those of E. Hopf, Ch. Morrey and L. Nirenberg.

the same time the existence of the derivatives of u. The construction of the expression proceeds in two parts:

1) Derivation of expressions for derivatives of weighted spherical means of u with respect to radius and center of the sphere.

2) Expressing u and its derivatives in terms of spherical means of u and their derivatives.

It will be proved that a continuous weak solution u of (1) can be differentiated any number of times and is a strict solution of (1), provided f and the coefficients of P possess a sufficient number of derivatives. Here u is called a weak solution of (1), if for every solid sphere S contained in the domain of definition of u

$$\int_S u\, M[v]\, dx_1 \ldots dx_n = \int_S fv\, dx_1 \ldots dx_n \tag{2}$$

for every v in C_∞ that vanishes in a neighborhood of the boundary of S. Here M denotes the adjoint differential operator to L.[4]

Let u satisfy (2) for all v of class C_∞ vanishing near the boundary of S. Then (2) will also hold for all v of class C_{2m} in S, which vanish with their derivatives of order $\leqq 2m$ on the boundary of S, since these more general v can be approximated properly by the special ones. Let S be a sphere of radius R and center $X = (X_1, \ldots, X_n)$. Then (2) holds in particular for all v of the form

$$v = (R-r)^a \cdot w(x, X, R)$$

where $a > 2m$, $r = |x-X|$, and w is an arbitrary function of class C_{2m} in x which vanishes identically for $r < \epsilon$. For $a \longrightarrow 2m+0$ the expression $M[v]$ is bounded in S and converges uniformly in every smaller concentric sphere to $M[(R-r)^{2m}w]$. Consequently

$$\int u\, M[(R-r)^{2m}w]\, dx_1 \ldots dx_n = \int f(R-r)^{2m} w\, dx_1 \ldots dx_n \tag{3}$$

Since $(R-r)^{2m}$ vanishes of order $2m$ on the boundary of S, we have

$$M[(R-r)^{2m}w]_{r=R} = w\, M[(R-r)^{2m}]_{r=R} = (2m)!\, wQ(x_1-X_1, \ldots, x_n-X_n)R^{2m}$$

where Q is the characteristic form of M, which is identical with that of L. Differentiating both sides of (3) with respect to R, we find then

[4] It is clear that any "strict" solution, i.e. any u of class C_{2m} satisfying (1), is a weak solution. It turns out to be just as easy to prove the existence of higher order derivatives for continuous weak solutions as for the class of strict solutions.

that

$$(2m)! \int_{r=R} uw\, Q(x-X)dS$$

(4)

$$= R^{-2m} \int_{r=R} \frac{\partial}{\partial R} \left(f\ (R-r)^{2m}w\right) - uM\left[\frac{\partial}{\partial R}\ (R-r)^{m}w\right]\ dx_1 \ldots dx_n$$

We now use the assumption that Q is definite, which implies that the left-hand side of the last equation represents an "arbitrary" weighted spherical mean of u. Let $q = q(x,X,R)$ be a function which is defined and of class C_s in x $(s \geqq 2m)$, and which together with its derivatives of order $\leqq s$ is analytic in R,X for $R > \epsilon$. We define the function w by

$$w(x, X, R) = \begin{cases} \dfrac{q(x,X,R)e^{-(R-\epsilon)/(r-\epsilon)}}{Q(x-X)} & \text{for } r = |x-X| > \epsilon \\ 0 & \text{for } r = |x-X| < \epsilon \end{cases}$$

Then w satisfies the same assumptions as q. The functions q^* and q^{**} shall be defined by

$$q^* = \frac{R^{-2m}}{(2m)!} \left(\frac{\partial}{\partial R}\ (R-r)^{2m}w\right)$$

and

$$q^{**} = \frac{R^{-2m}}{(2m)!} \frac{\partial}{\partial R}\ M[(R-r)^{2m}w]$$

Then identity (4) takes the form

$$(5)\quad \int_{r=R} u(x)q(x,X,R)dS = \int_{r<R} \left(f(x)q^*(x,X,R) - u(x)q^{**}(x,X,R)\right) dx_1 \ldots dx_n$$

where q^* and q^{**} satisfy the same regularity conditions as q, only with s replaced by $s - 2m$ in the case of q^{**}.

Formula (5) expresses a weighted surface integral of u in terms of weighted volume integrals of f and u. The volume integrals necessarily have first derivatives with respect to the radius R and the coordinates X_i of the center of the sphere; the same holds then for the surface integrals of u. One obtains the formulae

$$\frac{\partial}{\partial R}\int_{r=R} uq\,dS = \int_{r=R}(fq^{*}-uq^{**})dS + \int_{r<R}\left(f\frac{\partial q^{*}}{\partial R} - u\frac{\partial q^{**}}{\partial R}\right)dx_1\ldots dx_n$$

$$\frac{\partial}{\partial X_i}\int_{r=R} uq\,dS = \int_{r=R}(fq^{*}-uq^{**})\frac{x_i-X_i}{R}dS + \int_{r<R}\left(f\frac{\partial q^{*}}{\partial X_i} - u\frac{\partial q^{**}}{\partial X_i}\right)dx_1\ldots dx_n$$

These formulae express the first derivatives of a weighted spherical surface integral of u with respect to radius and center of the sphere in terms of surface and volume integrals of u and f. The formulae are valid whenever the derivatives of the weight q and of the coefficients of M and of f which occur in the expression exist and are continuous. But only continuity of u has been used. It is clear that we can iterate this process, and differentiate again, expressing by the same formulae the first derivatives of the surface integrals of u occurring on the right-hand side in terms of expressions not containing derivatives of u.

In this way one establishes the existence of and an expression for the derivatives of any order of a spherical average of u with weight q, provided that q, f and the coefficients of the differential equation can be differentiated sufficiently often.

In particular the function

$$I(X_1,\ldots,X_n,R) = I(X,R) = \int_{|x-X|=R} u(x)dS \tag{6}$$

will have derivatives of any order with respect to X and R for $R > 0$, provided u is a continuous weak solution of (1), and provided f and the coefficients of L have sufficiently many derivatives.

It will now be shown that for an odd number n of dimensions a function u can be expressed in terms of its spherical integrals $I(X,R)$ with $R > 0$, and that u can be differentiated any number of times, if I possesses sufficiently many derivatives.

The expression for u in terms of I that will be given here is based on the validity of Huygens' principle in the strong form, which is valid only for odd numbers of dimension. The case of an even n can then be handled easily by the method of descent. For simplicity in writing

I shall restrict myself to the case $n = 3$, the case of a general odd n not offering any additional difficulties.

Let $n = 3$. We consider the solution $U(x_1,x_2,x_3,t)$ of the wave equation

$$\Delta U = U_{x_1x_1} + U_{x_2x_2} + U_{x_3x_3} = U_{tt}$$

with initial data

$$U(x_1,x_2,x_3,0) = u(x_1,x_2,x_3), \quad U_t(x_1,x_2,x_3,0) = 0$$

For sufficiently regular functions u, e.g. polynomials, there exists a unique solution U which, by the classical formula of Poisson, is given by

(7) $$U(x,t) = \frac{\partial}{\partial t} \frac{1}{4\pi t} I(x,t)$$

where I is the spherical integral of u defined in (6)[5]. For every function $u(x_1,x_2,x_3)$ of class C_1 and every fixed positive t equation (7) defines a function U of x_1,x_2,x_3, if $I(x,t)$ denotes the integral of u over the sphere of radius t and center x. We write this function U symbolically in the form

(8a) $$U = c_t u$$

where $c_t u$ written out in detail is the function having at the point x the value

(8b) $$\frac{1}{4\pi t^2} \int (u+t\frac{du}{dN})dS = \frac{1}{4\pi t^2} \int (1+t \frac{d}{dN}) u \, dS$$

here the integral is extended over the sphere of radius t about the point x, and du/dN denotes the exterior normal derivative of u. Now symbolically the solution of the equation

$$U = U_{tt}$$

with initial values $U = u$, $U_t = 0$ can be written in the form

(9) $$U = \cos(it\sqrt{\Delta})u$$

[5] A similar formula involving derivatives of I of order $\leq \frac{n-1}{2}$ exists for every odd n. See Courant-Hilbert, Methoden d. Math. Physik II, p. 399.

suggesting the identification of the operator c_t with $\cos(it\sqrt{\Delta})$. The double angle formula for the cosine suggests then that

(10) $$1 = 2c_t c_t - c_{2t}$$

or that for a function $u(x)$

(11) $$u(x) = 2c_t c_t u - c_{2t} u$$

Using the expression (7) for $U = c_t u$ we obtain the formula

(12) $$u(X) = \frac{1}{8\pi^2 t^2} \int \left(1+t \frac{d}{dN}\right) \left(-\frac{I(x,t)}{t^2} + \frac{I'(x,t)}{t}\right) dS$$

$$+ \frac{1}{4\pi t^2} \left(\frac{I(x,2t)}{4t^2} - \frac{I'(x,2t)}{2t}\right)$$

where the integral is extended over the sphere of radius t about the point X in x-space, and $I'(x,t)$ is defined by $dI(x,t)/dt$. This is the desired expression for u in terms of $I(x,r)$ with positive r.

So far the derivation of (12) has been purely formal. However, the expression (9) for the solution of the wave equation is certainly valid when u is a polynomial, and hence expanding $\cos(it\sqrt{\Delta})$ into a power series will contribute only a finite number of terms. Thus the formula (12), which is based on (9), will be correct for polynomials u.

To prove (11) for more general u we introduce the operator s_t by

$$s_t u = \frac{1}{4\pi t} I(x,t)$$

For continuous u the expression $s_t u$ will be defined and continuous in x and t. Moreover, formally

$$\frac{d}{dt} s_t = c_t, \quad s_t = \int_0^t c_\xi \, d\xi$$

so that symbolically s_t can be written $(i\sqrt{\Delta})^{-1} \sin(it\sqrt{\Delta})$. For polynomials u we have the "addition theorem"

(13) $$2c_\beta s_\alpha u = s_{\alpha+\beta} u + s_{\alpha-\beta} u$$

Integrating this identity with respect to β, we obtain an identity free of derivatives of u. Since this identity is valid for all polynomials, it is valid for all continuous u.

Hence identity (13), which follows by differentiating with respect to β, must also be valid for continuous u, since the right-hand side is defined and continuous. Putting $\alpha = \beta + h$ we have then

$$2c_\beta s_{\beta+h} u = s_{2\beta+h}\, u + s_h u$$

and, in particular for $h = 0$,

$$2c_\beta s_\beta u = s_{2\beta} u$$

Hence, since $\lim_{h \to 0} h^{-1} s_h u = u$ for continuous u,

(14) $$u = \lim_{h \to 0} \; 2c_\beta \frac{s_{\beta+h} - s_\beta}{h} u - \frac{s_{2\beta+h} - s_2}{h} u$$

If now $I(x,t)$ has continuous second derivatives with respect to x and t for t positive and bounded away from 0, we can write (14) for positive β in the form

$$u = (2c_\beta c_\beta - c_{2\beta})u = 2c_\beta \frac{\partial}{\partial \beta}\left(\frac{I(x,\beta)}{4\pi\beta}\right) - \frac{\partial}{\partial 2\beta}\left(\frac{I(x,2\beta)}{8\pi\beta}\right)$$

which is equivalent to (11) or (12).

If the existence of 3rd derivatives of $I(x,t)$ for positive t has been established, we obtain the existence of first derivatives of u and an expression for these derivatives in the form

$$u_{x_i} = 2c_\beta \frac{\partial}{\partial \beta}\left(\frac{I_{x_i}(x,\beta)}{4\pi\beta}\right) - \frac{\partial}{\partial 2\beta}\left(\frac{I_{x_i}(x,2\beta)}{8\pi\beta}\right)$$

Continuing in this fashion we see that the existence of derivatives of order k for $I(x,t)$ implies the existence of derivatives of order $k - 2$ for u, in case $n = 3$. (Similarly for general odd $n > 1$ the fact that $I(x,t)$ is of class C_k for positive t will imply that u is at least of class C_{k-n+1}.)

In this way it follows that a function u, which is a continuous weak solution of an elliptic equation, will have any desired number of derivatives, if the coefficients of the equation are sufficiently often differentiable.

This type of argument applies only to the case of an odd number of dimensions, in which case the solutions of the wave equation involves only mean values of the initial data on spheres of positive radius. The analogous result for solutions of elliptic equations in an even number of dimensions can be obtained by the method of descent. A continuous weak solution $u(x_1, \ldots, x_n)$ of an elliptic equation (1) with positive definite characteristic form can also be considered as a continuous weak solution of the equation

$$(L+D_{n+1}^{2m})u = f(x_1, \ldots, x_n) \tag{15}$$

involving the additional independent variable x_{n+1}. If n is even, the preceding arguments can be applied to the equation (15), which is again elliptic in the $(n+1)$ coordinates $x_1, \ldots, x_{n+1}$. Regularity of u can then again be proved from that of f and of the coefficients of L.

BIBLIOGRAPHY

[1] BROWDER, F., "The Dirichlet problem for linear elliptic equations of arbitrary even order with variable coefficients," Proceedings of the National Academy of Sciences <u>38</u> (1952), 232.

[2] FRIEDRICHS, K. O., "On the differentiability of the solutions of linear elliptic differential equations," Communications on Pure and Applied Mathematics <u>6</u> (1953), 299-326.

[3] JOHN, F., "Derivatives of continuous weak solutions of linear elliptic equations," Communications on Pure and Applied Mathematics <u>6</u> (1953), 327-335.

[4] JOHN, F., "On linear partial differential equations with analytic coefficients," Communications on Pure and Applied Mathematics <u>2</u> (1949), 209-253.

[5] JOHN, F., "General properties of solutions of linear elliptic partial differential equations," Proceedings of the Symposium on Spectral Theory and Differential Problems (1951), 113-175.

[6] SCHWARTZ, L., Théorie des Distributions, Volume I, Hermann, Paris (1950).

IV. ON MULTIVALUED SOLUTIONS OF LINEAR PARTIAL DIFFERENTIAL EQUATIONS[1]

S. Bergman

§1

Multivalued solutions of linear partial differential equations

$$\Delta \Psi + F \Psi = 0 \tag{1}$$

in two and three variables play a role in various applications: e.g., in some problems in electricity; in the theory of compressible fluids (when we apply the hodograph method); etc.

For purposes of evaluation and investigation of singularities of these functions, it is of interest to obtain representations of solutions of (1) in the form

$$\Psi = \sum_{n=0}^{\infty} Q^{(n)} S^{(n)} \tag{2}$$

where the terms $Q^{(n)}$ depend only upon the equation and therefore can be tabulated once and for all, and $S^{(n)}$ can be represented in closed form using algebro-logarithmic expressions, Theta functions, their derivatives, and finitely many transcendental functions which are independent of the coefficient F of the equation. In the following we shall formulate some results pertaining to representations of this kind.

§2

Let

$$\Psi(\zeta, \bar{\zeta}) = \sum a_{mn} \zeta^{m} \bar{\zeta}^{n}, \quad \zeta = x + iy, \quad \bar{\zeta} = x - iy \tag{3}$$

be a real solution of a differential equation

[1] This work was done under a contract with the Office of Naval Research.

$$\Delta\Psi + F\Psi \equiv 4\frac{\partial^2\Psi}{\partial\zeta\partial\bar{\zeta}} + F\Psi = 0, \qquad \Delta \equiv \frac{\partial^2}{\partial x^2} + \frac{\partial^2}{\partial y^2} \tag{1a}$$

where F is an entire function of x and y (when continued to complex values of the variables).

Generalizing the operator "Re" (Real part of), one introduces to every differential equation (1a) operators

$$\mathrm{Re}[g(\zeta) + \sum_{n=1}^{\infty} Q^{(n)}(\zeta, \bar{\zeta})\, g_n(\zeta)] = \Psi(\zeta, \bar{\zeta}) \tag{2a}$$

$$g_n(\zeta) = \frac{(-1)^{n-1}}{2^{2n}B(n,n+1)} \int_0^{\zeta} (\zeta-\tau)^{n-1} g(\tau)\, d\tau$$

which transform analytic functions $g(\zeta)$ of one variable into solutions of (1a). ($B(n,n+1)$ are Beta functions.) See [4][2], p. 5.

If $Q^{(n)}(\zeta,0) = 0$, $n = 1,2,\ldots$, the operator (2a) is called integral operator of the first kind. (It should be mentioned that there exist various integral operators of the form (2a) which are not of the first kind.)

If

$$g(\zeta) = 2\Psi(\zeta,0) + \text{const.} \tag{4}$$

is an algebraic function of one complex variable defined on the Riemann surface of

$$A(\tau,\zeta) \equiv \sum_{\nu=0}^{n} A_\nu(\zeta)\tau^{n-\nu} = 0$$

where $A_\nu(\zeta)$ are polynomials, then Ψ can be written in the form (2a) where the $g_n(\zeta)$ $(n=1,2,\ldots)$ can be represented in a closed form using algebro-logarithmic functions, Theta functions, their derivatives, and finitely many transcendental functions, namely, integrals of the first kind defined on the Riemann surface of $A(\tau,\zeta) = 0$ (see [4], § 2 and § 3). The only singularities of $\Psi(\zeta, \bar{\zeta})$ are algebraic branch points and "pole-like" singularities of the first kind.[3] The location of these singularities

2 Numbers in brackets refer to bibliography.

3 It should be stressed that proceeding in a similar way using any integral operator $\{Q^{(n)}\}$ of the form (2a)(which does not satisfy necessarily the

depends only on $\{a_{mo}\}$ (m=1,2,...) but is independent of F. The singularities can be determined by means of theorems of Hadamard, Mandelbrojt, Szegö, Polya, Eisenstein, etc. if the subsequence $\{a_{mo}\}$ is given.

REMARK. We note that fundamental solutions

$$(5)\quad \psi^{(L)} = \frac{1}{2} A(\log s + \log \bar{s}) + B, \qquad s = \zeta - \zeta_0 \equiv (x-x_0) + i(y-y_0)$$

$$\bar{s} = \bar{\zeta} - \bar{\zeta}_0$$

which (for real values of x and y) are single-valued, are of great importance in various applications. (See [5] and [2], page 473, where explicit expressions for A and B are given.) Differentiating with respect to the parameter x_0 and y_0, we obtain fundamental solutions of higher order (see [2], page 474, Remark 5.1).

§3

The results of Section 2 can be generalized to the case of differential equations

$$(1.b)\qquad \frac{\partial^2 \psi}{\partial x^2} + \frac{\partial^2 \psi}{\partial y^2} + \frac{\partial^2 \psi}{\partial z^2} + F_1\psi = 0, \qquad \psi \equiv \psi(x,y,z)$$

in three variables, where $F_1 \equiv F_1(r^2)$ is an entire function (when continued to the complex value of the argument) of $r^2 = x^2 + y^2 + z^2$. In analogy to (2a) in this case, the formula (see [3], p. 500)

$$(2.b)\qquad \psi(x,y,z) = G(x,y,z) + \sum_{n=1}^{\infty} B^{(n)}(r^2)K_n(x,y,z)$$

$$K_n(x,y,z) = \int_0^1 (1-\sigma^2)^{n-1}\sigma^2 G(\sigma^2 x, \sigma^2 y, \sigma^2 z)\, d\sigma$$

holds where G is an arbitrary harmonic function of three variables which is regular at the origin. Here $B^{(n)}(r^2)$ are functions which depend only

---

condition $Q^{(n)}(\zeta,0) = 0$), we can generate solutions of (1a) possessing pole-like singularities belonging to the operator $\{Q^{(n)}\}$. We shall denote these singularities in the general case as pole-like singularities of type $\{Q^{(n)}\}$, while in the case where the $Q^{(n)}$'s yield the integral operator of the first kind we call them pole-like singularities of the first kind.

on F_1.

The formula (2.b) yields representations of multivalued solutions of (1.b), similar to those obtained in the two-dimensional case. If one chooses for G a rational harmonic function, e.g., $G(x,y,z) = (x-\alpha)[(x-\alpha)^2 + (y-\beta)^2]^{-1}$, then we obtain multivalued solutions of (1.b) possessing the property that the $K_n(x,y,z)$ are algebraic functions of x,y,z.

If one chooses for G an algebraic function[4] belonging to a field T, then one obtains classes of solutions which possess the property that every $K_n(x,y,z)$ can be represented in closed form by algebro-logarithmic functions, Theta functions, their derivatives, and finitely many transcendental functions.[5]

Real solutions Ψ of (1.b) which are regular at the origin can be developed in a sufficiently small neighborhood of the origin in the form

$$(6)\quad \Psi(\rho,\theta,\phi) = \sum_{n=0}^{\infty} \sum_{\nu=0}^{\infty} \rho^{n+\nu} \sum_{m=-n}^{n} A_m^{(m,\nu)} P_n^{|m|}(\cos\theta)\, e^{im\phi}$$

where $P_n^{|m|}$ are associated Legendre functions and $A_m^{(n,\nu)}$ are (complex) constants, $A_m^{(n,\nu)} = \overline{A_{-m}^{(n,\nu)}}$.

If the subsequence $[A_m^{(n,0)}]$ $(n = 0, 1, 2, \ldots, \ m = 0, 1, \ldots, n)$ satisfies certain conditions, see [4], p. 31, (7.8a) - (7.8c), and p. 25, of (6.4), then Ψ will admit the representation (2.b) with K_n representable in closed form by algebro-logarithmic expressions, Theta functions, their derivatives, and finitely many transcendental functions mentioned before. These functions and the singularity lines of Ψ are determined by the subsequence $[A_m^{(n,0)}]$ $(n=0,1,2,\ldots, \ m=0,1,\ldots,n)$ independently of F_1, see [4].

§4

If $\vec{V} = \{H_1, H_2, H_3\}$ where H_k are harmonic functions such that $\operatorname{curl} \vec{V} = 0$ and if j is a closed curve, various theorems can be derived stating how

$$(7)\qquad R = \int_j (H_1 dx + H_2 dy + H_3 dz)$$

depends upon the location of singularities of $\vec{V}$ and some properties of j.

[4] Classes of algebraic harmonic functions in three variables are considered in [3] Theorem 2.1, p. 475 and ff.

[5] These transcendental functions depend only upon T.

As can be shown, the formula (2.b) yields a representation of solutions of (1.b) in terms of harmonic functions H

$$\Psi(x,y,z) = P(H) \equiv \int_{-1}^{1} \Omega(r,\tau)H(x(1-\tau^2),y(1-\tau^2),z(1-\tau^2))d\tau \quad (2.c)$$

where Ω (the generating function of the operator) depends only on the equation, while H is connected with G by the relation

$$(8) \qquad G(x,y,z) = \int_{T=0}^{1} T^{-\frac{1}{2}} H(x(1-T),\ y(1-T),\ z(1-T))dT$$

See [1], p. 427, and [3], p. 500.

Thus, in particular if H is a rational harmonic function, G will be an algebraic harmonic function.

Let j be a closed (simple) curve which lies on the surface of a sphere with center at the origin, and let $\Psi_k = P(H_k)$, where H_k are components of $\vec{V}$. They are supposed to be regular at the origin.

Then, by (7)

$$\int_j (\Psi_1 dx + \Psi_2 dy + \Psi_3 dz)$$

$$= \int_{\tau=-1}^{1} \Omega(r,\tau) \int_j [H_1 dx + H_2 dy + H_3 dz] d\tau$$

$$(9) \qquad = \int_{\tau=-1}^{1} \Omega(r,\tau)R(\tau)d\tau,$$

$$H_k = H_k(x(1-\tau^2),\ y(1-\tau^2),\ z(1-\tau^2)),\ k = 1,\ 2,\ 3$$

(We note that $R(\tau)$, in general, depends on τ, since if τ varies, the arguments of H_k vary, and the value of R in (7) can change.)

Similar formulas can be obtained for algebraic harmonic functions H. It should be further mentioned that theorems of Abel's type for harmonic vectors, see [3], p. 497, lead to analogous theorems for the corresponding triple of solutions $\{\Psi_1, \Psi_2, \Psi_3\}$.

Finally, we wish to add that using representations (2), (2c), and similar formulas, various results in the theory of meromorphic functions and the theory of integrals of functions defined on Riemann surfaces of more general character yield relations for solutions of equation (1).

BIBLIOGRAPHY

[1] BERGMAN, S., "Classes of solutions of linear partial differential equations in three variables," Duke Mathematical Journal 13 (1946), 419-458.

[2] BERGMAN, S., "Two-dimensional subsonic flows of compressible fluids and their singularities," Transactions of the American Mathematical Society 62 (1947), 452-498.

[3] BERGMAN, S., "On solutions with algebraic character of linear partial differential equations," Transactions of the American Mathematical Society 68 (1950), 461-507.

[4] BERGMAN, S., "The coefficient problem in the theory of linear partial differential equations," Transactions of the American Mathematical Society 73 (1952), 1-34.

[5] SOMMERFELD, A., "Randwertaufgaben in der theorie der partiellen differentialgleichungen," Enzyklopädie der mathematischen Wissenschaften, Vol. II, 1, I.

V. FUNCTION-THEORETICAL PROPERTIES OF SOLUTIONS OF PARTIAL DIFFERENTIAL EQUATIONS OF ELLIPTIC TYPE[1]

L. Bers

§1. Introduction

In this paper we discuss function-theoretical properties of functions satisfying elliptic partial differential equations, without assuming the equations to be analytic.[2] We are concerned, in particular, with the unique continuation property of solutions, with isolated singularities, and with entire solutions (that is solutions defined for all values of the independent variables). While it would be natural to consider such questions for elliptic equations and systems of order $2m$ in n-space, the present state of knowledge necessitates the restriction to the simplest case: $m = 1$, $n = 2$.

For linear equations our results are rather comprehensive and confirm the belief that the case of the Laplace equation is "typical". The deeper reason for this analogy lies in the possibility of associating with every linear elliptic equation a generalized complex function theory (theory of pseudo-analytic functions[3]).

In this paper we do not give a complete description of this theory. In §2, however, we show how solutions of the general linear elliptic equation can be expressed in terms of pseudo-analytic functions, and in §3 we give a new proof of the important result of Carleman which asserts that the zeros of pseudo-analytic functions are isolated. §4 contains the statement and complete proof of the basic "similarity principle" for pseudo-analytic functions. This principle is then used for the construction of

---

[1] Some of the results reported here were obtained during work sponsored by the OOR.

[2] This degree of generality is not only interesting in itself but also indispensable for a harmonious in-the-large theory. On the other hand, it precludes the direct application of the methods of Bergman [3], [4] Vekua [29] and John [21].

[3] See [9], [10] and the papers by Položii, Petrovskii, Šabat, Lukomskaya, Bers and Gelbart, and others mentioned there. Cf. also the paper [1] by Agmon and the author, and a recent paper by Vekua [30].

the fundamental solution and Green's function (§5; this section serves only as an illustration since the results obtained are not new), for the proof of the topological equivalence between harmonic functions and solutions of linear elliptic equations (§6), and for a rather complete description of entire solutions (§7) and singularities (§8).

In the case of non-linear equations only fragmentary results are known. The results discussed in §§9-11 are patterned after the thoroughly explored case of minimal surfaces.[4] The equation of minimal surfaces is, however, "typical" for only one class of quasi-linear elliptic equations.

In §9 we show how solutions of certain quasi-linear equations can be expressed in terms of pseudo-analytic functions and describe a classification of such equations. §10 deals with extensions of Bernstein's theorem on minimal surfaces and §11 with those of the author's theorem on singularities of minimal surfaces. An application of pseudo-analytic functions to a non-linear boundary value problem is mentioned in §12. In view of limitations in space these results are stated without proofs.

NOTATION: We set $z = x + iy$, $\zeta = \xi + i\eta$ and denote complex conjugates by a bar. Functions of (x,y) will be written as functions of z, and we shall use the formal derivatives $w_z = (w_x - iw_y)/2$, $w_{\bar{z}} = (w_x + iw_y)/2$. A function $w(z)$, $z \in D$, will be called of class C if it is continuous in the domain D, of class H if it satisfies a uniform Hölder condition on every compact subset of D, of class C^n (or H^n) if it possesses partial derivatives of order n which are of class C (or H).

We note the well known and easily verified[5] identity

$$(1.1) \qquad w(z) = \frac{1}{2\pi i} \oint_{D'} \frac{w(\zeta)\, d\zeta}{\zeta - z} - \frac{1}{\pi} \iint_D \frac{w_{\bar{\zeta}}(\zeta)\, d\xi\, d\eta}{\zeta - z}, \qquad z \in D,$$

which holds for every function which is of class C^1 in a bounded domain D with sufficiently smooth boundary D', of class C on $D \cup D'$, and such that the double integral converges absolutely.

If $\rho(z)$, $z \in D$, is a measurable function, we set

$$(1.2) \qquad I_D(\rho \,\|\, z) = -\frac{1}{\pi} \iint_D \frac{\rho(\zeta)\, d\xi\, d\eta}{\zeta - z}.$$

A non-negative function $K(z)$ will be called <u>admissible</u> in D if the inequality $|\rho(z)| \leq K(z)$ implies that

---

[4] Cf. the survey in [6].

[5] Apply Green's theorem to the line integral in (1.1).

$$|I_D(\rho \| z)| \leq M(1+|z|)^{-1-\varepsilon}, \quad |I_D(\rho \| z_1)-I_D(\rho \| z_2)| \leq M|z_1-z_2|^{\varepsilon},$$

M and ε being positive constants depending only on K.

LEMMA 1. Let $\Gamma_1, \ldots, \Gamma_k$ be distinct simple closed continuously differentiable curves, and denote by $\delta_j(z)$ the distance from z to Γ_j. If $K(z) \geq 0$, $K(z) = O(|z|^{-1-\varepsilon})$, $z \longrightarrow \infty$, $K(z) = O(\delta_j(z)^{\varepsilon-1})$, $z \longrightarrow \Gamma_j$, $j = 1, \ldots, k$, $\varepsilon > 0$, then $K(z)$ is admissible (in the whole plane).

We omit the proof which proceeds along standard lines.

LEMMA 2. Let $\rho(z)$ be admissible in a domain D. If $\rho(z)$ is of class H in a subdomain $D_0 \subset D$, then $I_D(\rho \| z)$ is of class H^1 in D_0 and $\partial I_D(\rho \| z)/\partial \bar{z} = \rho(z)$.

This is a restatement of a classical result in the theory of the logarithmic potential.[6]

§2. Linear Elliptic Equations and Pseudo-analytic Functions

A linear elliptic equation

$$\begin{aligned} &A_{11}(x,y)\phi_{xx} + 2A_{12}(x,y)\phi_{xy} + A_{22}(x,y)\phi_{yy} \\ (2.1)\qquad &+ A_1(x,y)\phi_x + A_2(x,y)\phi_y + A_0(x,y)\phi = 0, \end{aligned}$$

can be brought, by introducing new dependent and independent variables, into the canonical form

$$(2.2) \qquad \phi_{xx} + \phi_{yy} + \alpha(x,y)\phi_x + \beta(x,y)\phi_y = 0.$$

This transformation[7] is possible in every domain in which the A_{ij} are

[6] Note that $I_D(\rho\|z)$ is the z-derivative of a logarithmic potential.

[7] It consists of introducing the new unknown function ϕ/ϕ_0 (ϕ_0 being a fixed positive solution) and new independent variables conformal with respect to the metric $A_{22}dx^2 - 2A_{12}dx\,dy + A_{11}dy^2$.

of class H^1 and (2.1) possesses a positive solution. From now on we consider only equation (2.2) assuming[8] once and for all that α and β are of class H.

Equation (2.2) can be rewritten in the form of a system in two different ways. Set

$$u = \phi_x, \quad v = -\phi_y. \tag{2.3}$$

If ϕ satisfies (2.2), then u and v satisfy the equations

$$\begin{aligned} u_x - v_y &= -\alpha u + \beta v, \\ u_y + v_x &= 0, \end{aligned} \tag{2.4}$$

and if u and v satisfy (2.4), then relations (2.3) determine a (perhaps multiple-valued) solution of (2.2).

On the other hand, assume that we found functions $\sigma(x,y) > 0$, $\tau(x,y)$ of class H^1 such that

$$\sigma_x - \tau_y = \alpha\sigma, \quad -\sigma_y + \tau_x = \beta\sigma. \tag{2.5}$$

If ϕ is a solution of (2.2), then there exists a "conjugate" (perhaps multiple-valued) function $\psi(x,y)$ such that

$$\begin{aligned} \sigma\phi_x + \tau\phi_y &= \psi_y, \\ -\tau\phi_x + \sigma\phi_y &= -\psi_x. \end{aligned} \tag{2.6}$$

Conversely, if (ϕ,ψ) satisfy (2.6), then ϕ satisfies[9] (2.2).

Now let a(z), b(z) be two (in general, complex-valued) functions of class H. A function w(z) of class C^1 is called [a,b] pseudo-analytic (of the first kind) if

$$w_{\bar{z}} = aw + b\bar{w}\ . \tag{2.7}$$

Noting relations (2.3), (2.4) we obtain the following lemma.

---

[8] This assumption could be weakened for many purposes.

[9] It is known and not difficult to prove that solutions of (2.6) must be of class H^2.

LEMMA 3. Set

$$4a_1(z) = -\alpha(z) - i\beta(z). \tag{2.8}$$

If ϕ is a solution of (2.2), then $w = \phi_x - i\phi_y$ is an $[a_1, \overline{a}_1]$ pseudo-analytic function. Conversely, if $w(z)$ is $[a_1, \overline{a}_1]$ pseudo-analytic, then

$$\phi(z) = \text{Re} \left\{ \int_{z_0}^{z} w \, dz \right\}$$

is a (perhaps multiply-valued) solution of (2.2).

Two $[a,b]$ pseudo-analytic functions $F(z)$, $G(z)$ defined in a domain D are said to form a generating pair (F,G) belonging to $[a,b]$ if

$$\text{Im}\left\{\overline{F}G\right\} > 0. \tag{2.9}$$

Clearly, any two functions $F(z)$, $G(z)$ of class H^1 satisfying (2.9) form a generating pair belonging to some uniquely determined $[a,b]$. A generating pair is called bounded if F, G are bounded and $\text{Im}\{\overline{F}G\}$ bounded from below.

LEMMA 4. (i) If $a(z)$, $b(z)$ are functions of class H and $|a| + |b|$ is admissible in D, then there exists in D a bounded generating pair (F, G) belonging to $[a,b]$.

(ii) If $a = b$, (F,G) may be chosen so that $G \equiv 1$.

The proof will be given in §4.

Inequality (2.9) implies that every function $w(z)$ defined in a subdomain of D admits the unique representation

$$w(z) = \phi(z)F(z) + \psi(z)G(z) \tag{2.10}$$

with real ϕ, ψ. Assume that ϕ and ψ are of class C^1. Since

$$F_{\overline{z}} = aF + b\overline{F}, \quad G_{\overline{z}} = aG + b\overline{G}$$

we have that $w_{\overline{z}} = aw + b\overline{w} + F\phi_{\overline{z}} + G\psi_{\overline{z}}$. Hence $w(z)$ is $[a,b]$

pseudo-analytic if and only if

$$F\phi_{\bar z} + G\psi_{\bar z} = 0. \tag{2.11}$$

If this condition is satisfied we say that $\omega(z) = \phi + i\psi$ is an (F,G) pseudo-analytic function of the <u>second kind</u>, and w an (F,G) pseudo-analytic of the first kind.

LEMMA 5. Let $|\alpha(z)| + |\beta(z)|$ be admissible in a domain D. Then there exist in D bounded real-valued functions σ, τ of class H^1 such that σ is positive and bounded away from zero, and (σ, τ) satisfy (2.5). If (F,G) are defined by setting

$$F = 1 - \frac{i\tau}{\sigma}, \qquad G = \frac{i}{\sigma}, \tag{2.12}$$

then every solution ϕ of (2.2) is the real part of an (F,G) pseudo-analytic function of the second kind (and of an (F,G) pseudo-analytic function of the first kind), and vice versa.

PROOF. According to Lemma 4 there exists a bounded generating pair $(\sigma+i\tau, i)$ belonging to $[-a_1, -a_1]$, a_1 being defined by (2.8). A direct computation verifies that σ and τ have the desired properties, and that equation (2.11), with F and G given by (2.12), is equivalent to the system (2.6). Hence ϕ is a solution of (2.2) if and only if there exists a ψ such that the function defined by (2.10) is (F,G) pseudo-analytic of the first kind. The corresponding function of the second kind, $\phi + i\psi$, has the same real part as w for $\operatorname{Re}\{F\} = 1$, $\operatorname{Re}\{G\} = 0$.

If the function (2.10) is pseudo-analytic, the function

$$\dot{w}(z) = \phi_{\bar z} F + \psi_{\bar z} G$$

is called the (F,G)-derivative of w. If (F,G) are given by (2.12), then $\dot{w} = \phi_x - i\phi_y$, so that $\dot{w}$ is pseudo-analytic by virtue of Lemma 3, though with respect to different coefficients $[a,b]$ than w.

It is easy to check that

$$\dot{w}(z_0) = \lim_{z \to z_0} \frac{w(z) - \phi(z_0)F(z) - \psi(z_0)G(z)}{z - z_0}$$

In the general theory of pseudo-analytic functions, pseudo-analyticity is defined by the existence of this limit in a domain, and it is shown that to every generating pair (F_0, G_0) there exists a sequence of generating pairs $\{(F_\nu, G_\nu)\}$ such that the (F_ν, G_ν)-derivatives of (F_ν, G_ν) pseudo-analytic functions are $(F_{\nu+1}, G_{\nu+1})$ pseudo-analytic, $\nu = 0, \pm 1, \pm 2, \ldots$.

In what follows we shall need one more way of defining pseudo-analyticity.

LEMMA 6. Let $a(z)$, $b(z)$ be of class H and $|a| + |b|$ admissible in D. A continuous function $w(z)$ defined in a subdomain $\Delta \subset D$ is $[a,b]$ pseudo-analytic of the first kind if and only if the function

$$h(z) = w(z) - I_\Delta(aw + b\overline{w} \| z)$$

is analytic in Δ.

PROOF. If w is pseudo-analytic, then by Lemma 2, $\partial I_\Delta/\partial \overline{z} = aw + b\overline{w}$, so that $h_{\overline{z}} \equiv 0$ and h is analytic. If h is analytic, then w is of class H, so that again $\partial I_\Delta/\partial \overline{z} = aw + b\overline{w}$, and since $h_{\overline{z}} = 0$, $w_{\overline{z}} = aw + b\overline{w}$.

The following statements are almost immediate consequences of Lemma 6. The class of pseudo-analytic functions is closed under <u>bounded convergence</u>. A uniformly bounded sequence of pseudo-analytic functions contains a <u>convergent subsequence</u>. An isolated singularity of a pseudo-analytic function is <u>removable</u>, if the function is single-valued and bounded.

§3. Carleman's Theorem

If the coefficients of (2.2) are analytic, the solutions are also analytic and hence possess the <u>unique continuation property</u>: they are uniquely determined by their values in any open set. Hadamard conjectured[10] and Carleman proved that the unique continuation property for equation (2.2), and hence also for equation (2.1), is independent of the analyticity hypothesis. The corresponding problem for equations with more than two independent variables is still open.

---

[10] In his book [18].

In view of Lemma 3 the unique continuation property is an immediate consequence of

THEOREM 1. (Carleman[11]). Let $w(z)$ be a function of class C^1 satisfying equation (2.7) with bounded a, b. If $w(z) \not\equiv 0$, then the zeros of $w(z)$ are discrete (and not of infinite order).

We give here a new proof of Carleman's theorem. We may assume that $w(z)$ is bounded and defined in a bounded domain D, and that the (open) set $\Sigma \subset D$ on which $w \neq 0$ is not empty. Set (cf. (1.2))

$$s(z) = I_{\Sigma}\left(a + b\frac{\overline{w}}{w} \parallel z\right), \qquad f(z) = e^{-s(z)}w(z),$$

and note that $s(z)$ is bounded. Let z_0 be an accumulation point of $D - \Sigma$. Then there is a sequence of points z_ν in D with $z_\nu \longrightarrow z_0$, $w(z_\nu) = 0$. By a simple compactness argument and by the mean value theorem we conclude that for some real θ, $w_x(z_0)\cos\theta + w_y(z_0)\sin\theta = 0$. Since $w(z_0) = 0$, we also have by (2.7) that $w_x(z_0) + iw_y(z_0) = 0$. Hence $w_x(z_0) = w_y(z_0) = 0$, $w(z)$ and hence also $f(z)$ are $o(|z-z_0|)$ for $z \longrightarrow z_0$ and $f'(z_0)$ exists (and equals zero).

Next, let Δ be a disc in Σ. $s(z)$ differs from $I_\Delta(a + b\overline{w}/w \parallel z)$ by a function which is analytic in Δ, and I_Δ is of class C^1 in Δ and $\partial I_\Delta/\partial\overline{z} = a + b\overline{w}/w$. The last statement follows by applying to $\log w$ and the domain Δ the identity (1.1). We conclude that $f(z)$ is of class C^1 in Σ and that $f_{\overline{z}} = e^{-s}(-ws_{\overline{z}} + w_{\overline{z}}) = 0$. Thus $f(z)$ is analytic in D, except perhaps at the isolated points of $D - \Sigma$. It is also analytic at these points since it is bounded. Analyticity of f and boundedness of s imply the assertion.

The unique continuation property implies that the Cauchy problem from equation (2.2) has at most one solution. Let ϕ be a solution of (2.2) defined on one side of a continuously differentiable arc Γ and vanishing on Γ together with its normal derivative. Defining ϕ to be 0 on the other side of Γ we obtain a so-called weak solution[12] which vanishes on a domain. Since weak solutions are known to be ordinary solutions, we may conclude that $\phi \equiv 0$. Since the unique continuation property is obviously implied by the uniqueness theorem for the Cauchy problem,

---

[11] Carleman [14]. The smoothness condition of w can be weakened. An analysis of Carleman's proof was given by Miškis [25], a practically equivalent theorem by Dressel and Gergen [15].

[12] Concerning weak solutions see, for instance, John [22], and the papers by Browder and Friedrichs quoted there.

it is equivalent to it (and was conjectured by Hadamard in this form). Actually our proof implies a stronger result.

THEOREM 2. Let ϕ be a solution of (2.2) defined on one side of a continuously differentiable curve Γ. If $\phi_x, \phi_y \longrightarrow 0$ as (x,y) approaches non-tangentially any point of a set γ on Γ, and if the linear measure of γ is positive, $\phi \equiv 0$.

PROOF. By Lemma 3 and the proof of Theorem 1, we have $\phi_x - i\phi_y = e^s f$, where s is bounded and f analytic. The assertion follows by applying to f the classical theorem of Privaloff.

The extension of the unique continuation property from linear to certain classes of non-linear equations of elliptic type, for instance to equations considered in §9 below, proceeds by standard methods.

§4. Similarity Principle

We state now a general structure and existence theorem which has many applications.

Two functions, $f(z)$ and $w(z)$, defined in a domain D, will be called similar if $w(z) = e^{s(z)}f(z)$, $s(z)$ being continuous on the closure of D, and at $z = \infty$ if D is unbounded.

THEOREM 3. (Similarity Principle). Let $a(z)$, $b(z)$ be functions of class H defined in a domain D, and let $|a(z)| + |b(z)|$ be admissible in D. Every $[a,b]$ pseudo-analytic function $w(z)$ in D is similar to an analytic function $f(z)$, and every analytic function $f(z)$ is similar to an $[a,b]$ pseudo-analytic function $w(z)$. If one of the functions is given, the other may be chosen so that the ratio $(w/f) = S(z)$ satisfies the following conditions:

(i) $1/K \leq |S(z)| \leq K$, where $K > 1$ and depends only on a, b and D,

(ii) $|S(z_1)-S(z_2)| \leq k|z_1-z_2|^{\varepsilon}$, where k and ε are positive and depend only on a, b and D.

(iii) $|S(z_1)| = 1$ at a given point z_1 in the closure of D (or at infinity if D is unbounded),

(iv) $\operatorname{Im}\{S(z)\} = 0$ either (α) at a given point z_2 chosen in the same way as z_1, or (β) on a simple closed continuously differentiable curve Γ consisting of boundary points of D and containing D either in its interior or in its exterior. (Condition (β) is meaninful only if such a curve exists.)

PROOF. We treat first the case (α). Assume that $w \neq 0$ is given and set

$$s_0(z) = I_D(a+b\overline{w}/w \,\|\, z),$$

(4.1)

$$s(z) = s_0(z) - \operatorname{Re}\{s(z_1)\} - i \operatorname{Im}\{s(z_2)\}$$

The definition of s_0 is legitimate since the zeros of w are isolated. By the definition of admissibility $s_0(z)$ and $s(z)$ have a bound independent of w and satisfy a uniform Hölder condition independent of w. Reasoning as in the proof of Theorem 1 we conclude that $f = e^{-s}w$ is the desired function.

Assume now that f is given. Let B be the real Banach space of complex-valued bounded continuous functions $\sigma(z)$, $z \in D$, with the norm $\|\sigma\| = \text{l.u.b.}\, |\sigma(z)|$. For every $\sigma \in B$ set

$$\tau(z) = I_D(a + be^{\overline{\sigma}-\sigma}\,\overline{f}/f \,\|\, z) \tag{4.2}$$

Again we can give a bound for $|\tau|$ and a uniform Hölder condition for $\tau(z)$ which do not depend on σ or f. Define the operator $\hat{\sigma} = T\sigma$ by the relation

$$\hat{\sigma}(z) = \tau(z) - \operatorname{Re}\{\tau(z_1)\} - i \operatorname{Im}\{\tau(z_2)\} \tag{4.3}$$

Since T is continuous and a completely continuous[13] mapping of B onto a bounded subset of B, it follows from the Birkhoff-Kellogg-Schauder fixed point theorem that the equation $\sigma = T\sigma$ has a solution[14]. Since this σ differs from τ by an analytic function, σ is of class H. Hence, by Lemma 2, σ is of class H^1 except at the zeros of f, and $\sigma_{\overline{z}} = a + be^{\overline{\sigma}-\sigma}\,\overline{f}/f$. Set $w = e^{\sigma}f$. Then w is of class H^1, $w_{\overline{z}} = aw + b\overline{w}$

[13] The easy verification of the continuity of T may be left to the reader.

[14] Another arrangement of the proof avoids the use of non-linear equations and fixed point theorems.

except at the zeros of f. Noting the removable singularity theorem stated at the end of §2 we conclude that w is the desired function.

In case (β), let Δ be the simply connected domain bounded by Γ which contains D. For every function $h(z)$ which is bounded and continuous on the closure of D and satisfies on Γ a uniform Hölder condition there exists in Δ an analytic function $h^*(z)$ continuous on Γ and satisfying the conditions: $\operatorname{Im}\{h^*(z)\} = \operatorname{Im}\{h(z)\}$, $z \in \Gamma$, and $\operatorname{Re}\{h^*(z_1)\} = \operatorname{Re}\{h(z_1)\}$. This follows from Privaloff's theorem on conjugate functions which also implies a uniform Hölder condition for $h^*(z)$ on $\Delta \cup \Gamma$. We can repeat the proof given above replacing the definitions of $s(z)$ in (4.1) and of $\hat{\sigma}(z)$ in (4.3) by the equations: $s = s_0 - s_0^*$, $\hat{\sigma} = \tau - \tau^*$.

We are now in a position to prove Lemma 4 of §2.

Assume first that D is the whole plane, and denote by $W_\theta(z)$ the $[a,b]$ pseudo-analytic function similar to $f(z) = \text{const} = e^{i\theta}$ and such that $W_\theta(\infty) = e^{i\theta}$. This function is uniquely determined, for if $\widetilde{W}_\theta(z)$ were another such function, the difference $W_\theta - \widetilde{W}_\theta$ would be similar to an entire analytic function vanishing at infinity. Set $F = W_0$, $G = W_{\pi/2}$. Then $F_\theta = \cos\theta\, F + \sin\theta\, G$ and by Theorem 3 (i)

$$0 < 1/K \leq |\cos\theta\, F(z) + \sin\theta\, G(z)| \leq K, \quad 0 \leq \theta \leq 2\pi,$$

where K depends only on $[a,b]$. This implies the existence of a constant M such that

$$(4.4) \qquad |F(z)|,\ |G(z)| \leq M, \quad \operatorname{Im}\{\overline{F(z)}G(z)\} \geq 1/M > 0.$$

(F,G) is the desired generating pair. To prove statement (ii) observe that if $a = b$, then $G = i$.

If D is not the whole plane, we can find a sequence of pairs of functions of class H, $\{[a_\nu(z), b_\nu(z)]\}$, such that $a_\nu(z)$, $b_\nu(z)$ are defined for all z and vanish outside D; and such that for $|z| > R_\nu \longrightarrow +\infty$, $|a_\nu(z)| \leq |a(z)|$, $|b_\nu(z)| \leq |b(z)|$ for all z in D, $a_\nu(z) \equiv a(z)$, $b_\nu(z) \equiv b(z)$ for $z \in D_\nu$, D_ν being a subdomain of D, $D_\nu \longrightarrow D$. For each ν we can find a generating pair (F_ν, G_ν) belonging to $[a_\nu, b_\nu]$ and satisfying (4.4), M being independent of ν. From the convergence theorems stated at the end of §2 we conclude that a subsequence of $\{(F_\nu, G_\nu)\}$ converges in D to a desired generating pair (F,G). Under the hypothesis of (ii) all $G_\nu \equiv i$, so that $G \equiv i$.

§5. Fundamental Solution, Green's Function, Dirichlet Problem

In order to illustrate the power of the similarity principle we consider here some classical problems for equation (2.2), which could also be treated by other methods. We assume the coefficients α, β to be defined and $|\alpha| + |\beta|$ to be admissible in a simply connected domain D. Let a_1 denote the function defined by (2.8).

Assume first that D is the whole plane. Let $w(z)$ be the $[a_1, \bar{a}_1]$ pseudo-analytic function similar to $f = 1/(z-z_0)$ and such that $w/f = 1$ at $z = z_0$.

THEOREM 4. *The function*

$$\Lambda(z_1;z) = \text{Re}\left\{\int_{z_0+1}^{z} w\, dz\right\}$$

is a fundamental solution of (2.2) defined in the whole plane with singularity at $z = z_0$.

PROOF. We note Lemma 4 and observe that, by Theorem 3 (ii), $w(z) = (z-z_0)^{-1} + O(|z-z_0|^{\varepsilon-1})$, $z \longrightarrow z_0$. This implies that Λ is single-valued and has the right (logarithmic) singularity at z_0.

Now let D be bounded and let its boundary be a twice continuously differentiable curve Γ. (Note that α and β may become infinite on Γ). Let $g(z_0;z)$ be the Green's function for the Laplace equation and the domain D with the singularity at z_0, and let w be the $[a_1, \bar{a}_1]$ pseudo-analytic function similar to $f = g_x - ig_y$ and such that $|w/f| = 1$ at z_0 and $\text{Im}\{w/f\} = 0$ on Γ. Let z_1 be some point on Γ.

THEOREM 5. *The function*

$$G(z_0;z) = \text{Re}\left\{\int_{z_1}^{z} w\, dz\right\}$$

is the Green's function for the equation (2.2) and the domain D, with the singularity at z_0.

PROOF. Noting Lemma 3, we observe that G is single-valued. Indeed, $g = 0$ and hence $g_x dx + g_y dy = 0$ on Γ, so that

$$\text{Re}\left\{\oint_{\Gamma} w\, dz\right\} = 0.$$

The same argument shows that $g = 0$ on Γ. By Theorem 3 (ii) and by construction $w(z) = e^{i\gamma}(z-z_o)^{-1} + O(|z-z_o|^{\varepsilon-1})$, $z \longrightarrow z_o$, γ real. Since G is single-valued, $\gamma = 0$, and this implies that $G(z;z_o) - \log|z-z_o|$ is continuous at $z = z_o$.

Note that we proved not only the existence of Green's function, but also the existence and continuity of its <u>normal derivative</u>.

Let γ be an arc of Γ with the endpoints z_1, z_2, $\tilde{\gamma}$ the complementary arc, z_3 a point on $\tilde{\gamma}$. Let $m(\gamma;z)$ denote the harmonic measure of γ, that is the bounded harmonic function which equals 1 on γ and 0 on $\tilde{\gamma}$. Let w be the $[a_1,\overline{a}_1]$ pseudo-analytic function similar to $f = m_x - im_y$, and such that f/g is real on Γ and $|w/f| = 1$ at z_1.

THEOREM 6. (i) The function

$$M(\gamma;z) = \operatorname{Re}\left\{\int_{z_3}^{z} w\,dz\right\}$$

is a bounded solution of (2.2) which vanishes on $\tilde{\gamma}$ and equals 1 on γ.

(ii) There exists a constant K depending only on a,b such that

$$m(\gamma;z)/K \leq M(\gamma;z) \leq Km(\gamma;z). \tag{5.1}$$

The proof of (i) follows the same pattern as before and may be omitted. We remark only that it is based on the relations $m_x - im_y = t_j(z-z_j)^{-1} + O(|z-z_j|^{\varepsilon-1})$ $z \longrightarrow z_j$, $t_j \neq 0$, $\varepsilon > 0$, $j = 1,2$ which are easily established, say by mapping D conformally onto the unit disc. In order to prove (ii) let z be a fixed point in D. Since $m_x - im_y \neq 0$ in D there exists a smooth curve C joining a point z' on $\tilde{\gamma}$ to z such that $(m_x-im_y)(dx+idy)$ is positive on C. Integrating along C and noting Theorem 3 (i) we have that

$$M(\gamma;z) = \int_{z'}^{z} \operatorname{Re}\left\{w\,dz\right\} \leq K \int_{z'}^{z} m_x dx + m_y dy = K\,m(\gamma;z).$$

The second inequality in (5.1) can be established similarly.

Theorem 6 implies the solvability of the <u>Dirichlet problem</u> for equation (2.2). One can also obtain an integral representation for all <u>positive solutions</u> of (2.2), Fatou's theorem for such solutions, etc.

§6. Topological Equivalence

Next, we prove

THEOREM 7. Let $\phi(x,y)$ be a single-valued solution of (2.2) which may possess in its domain of definition, D, isolated singularities. Let $|\alpha| + |\beta|$ be admissible in D. There exists a homeomorphism $\zeta = \chi(z)$ of D onto a plane domain Δ which takes ϕ into a harmonic function (with isolated singularities).

Let σ, τ, F, G be the functions described in Lemma 5. If Σ denotes the discrete set of singularities of ϕ, then in $D - \Sigma$ there exists an (F,G) pseudo-analytic function of the second kind $\omega = \phi + i\psi$. ϕ and ψ satisfy equations (2.6), so that

$$J = \phi_x\psi_y - \phi_y\psi_x = \sigma(\phi_x^2 + \phi_y^2).$$

If ϕ is not a constant, then by Lemma 3 and Theorem 1, $J > 0$ except on a discrete set Σ_0. In $D_0 = D - (\Sigma \cup \Sigma_0)$ we define the Riemannian metric

$$d\phi^2 + d\psi^2 = g_{11}dx^2 + 2g_{12}dx\,dy + g_{22}dy^2, \tag{6.1}$$

$$g_{11}g_{22} - g_{12}^2 > 0.$$

The eccentricity of this metric equals

$$E = \frac{g_{11}+g_{22}}{2\sqrt{g_{11}g_{22}-g_{12}^2}} = \frac{1+\sigma^2+\tau^2}{2\sigma}$$

and is bounded by Lemma 5. There exists[15] a homeomorphism $\zeta = \chi(z)$ of D_0 onto a plane domain Δ_0 which is conformal with respect to the metric (6.1). In Δ_0 the function $\phi + i\psi$ is regular analytic. Now we recall that a mapping of bounded eccentricity preserves the non-degeneracy of boundary components[16]. Hence if z_0 is a point of $\Sigma \cup \Sigma_0$, the boundary

[15] In view of the general uniformization theorem.

[16] Because a mapping of bounded eccentricity cannot take a non-degenerate annulus onto a punctured disc. (There is, unfortunately, no comprehensive report on the theory of quasi-conformal mappings. For a good introduction and several references see the recent paper by Pfluger [26]).

continuum of Δ corresponding to z_0 must be a point. Thus χ may be extended to a homeomorphism of D onto a plane domain Δ .

§7. Entire Solutions of Linear Equations

Let the coefficients α, β of (2.2) be defined in the whole plane. Theorem 7 gives a complete topological description of all possible entire solutions of (2.2). Under the assumption (A): $|\alpha| + |\beta|$ admissible in the whole plane, it follows from Lemma 3 and the similarity principle that (non-constant) entire solutions actually exist. It also follows, in view of Lemma 5, that there exist (over the whole plane) a bounded generating pair (F,G) of the form (2.12) such that every solution of (2.2) is the real part of an (F,G) pseudo-analytic function of the first (or second) kind. (F,G) belongs to a definite coefficient pair [a,b]. We assume now that (B): $|a| + |b|$ is admissible in the whole plane. It can be shown[17] that condition (B) is satisfied if $|\alpha(\zeta)| + |\beta(\zeta)| = O(|z|^{-1-\varepsilon})$, $z \longrightarrow \infty$, and if for $|z| > R$ the functions $\alpha(z)$, $\beta(z)$ satisfy a Hölder condition with a fixed exponent η and constant $CR^{-1-\varepsilon}$ ($\varepsilon > 0$, $0 < \eta < 1$, $C > 0$).

For every integer n and for every complex c let $w(z) = Z^{(n)}(c,z_0;z)$ be the [a,b] pseudo-analytic function similar to $f(z) = a(z-z_0)^n$ and such that $w/f = 1$ at $z = z_0$. It is not difficult to see that the functions $Z^{(n)}$ ("global formal powers") are determined uniquely.

Set $\Phi_n(z_0;z) = \operatorname{Re}\{Z^{(n)}(1,z_0;z)\}$, $\tilde{\Phi}_n(z_0;z) = \operatorname{Re}\{Z^{(n)}(i,z_0;z)\}$, $\Phi_n(z) = \Phi_n(0;z)$, $\tilde{\Phi}_n(z) = \tilde{\Phi}_n(0;z)$. For $n = 1,2,\ldots,$ the functions $\Phi_n(z)$, $\tilde{\Phi}_n(z)$ are entire solutions of (2.2) "corresponding" to the harmonic polynomials $r^n \cos n\theta$, $r^n \sin n\theta$ ($z = re^{i\theta}$). From the expansion theorem for pseudo-analytic functions[18] we obtain

THEOREM 8. Every entire solution ϕ of (2.2) admits the unique representation

$$\phi(z) = \sum_{n=0}^{\infty} A_n \Phi_n(z) + B_n \tilde{\Phi}_n(z) \tag{7.1}$$

with real A_n, B_n valid in the whole plane. Conversely, if $\limsup |A_n + iB_n|^{1/n} = 0$, the series (7.1) converges and represents an entire solution.

---

[17] See [10], §15.

[18] See the paper [1] by Agmon and the author for the statement and proof of this theorem.

Consider now equation (2.2) without making <u>any</u> assumptions on the behavior of the coefficients at infinity. Or, more generally, consider an elliptic equation (2.1), with $A_0 \equiv 0$, in a domain D without assuming anything about the behavior of the coefficients at the boundary. Does there exist a single non-constant solution in the whole plane (in the whole domain D)? This question is still open. It can be answered, however, in the case of a <u>system</u> of two first order equations.

Let the coefficients of the system

$$\phi_x = A_{11}(x,y)\psi_x + A_{12}(x,y)\psi_y,$$

$$(7.2) \qquad \phi_y = A_{21}(x,y)\psi_x + A_{22}(x,y)\psi_y,$$

be defined and of class H^1 in a domain D, which may be the whole plane. We assume that in D

$$4A_{12}A_{21} + (A_{11}-A_{22})^2 < 0, \qquad A_{12} > 0,$$

(ellipticity condition) but impose no restrictions on the growth of the coefficients.

THEOREM 9. There exists in D a solution (ϕ,ψ) of system (7.2) such that $\phi_x\psi_y - \phi_y\psi_x > 0$ and the mapping $(x,y) \longrightarrow (\phi,\psi)$ is a homeomorphism of D. The solution may be chosen so that $\phi = \psi = \phi_x = 0$, $\phi_y = 1$ at a given point of D.

The proof of this theorem will be published elsewhere.[19]

§8. Singularities of Solutions of Linear Equations

We consider now isolated singularities of solutions of an elliptic equation (2.1). Since the problem is of a local character we lose no generality in assuming the equation to be in the form (2.2), with conditions (A), (B) of §7 satisfied. Let [a,b] have the same meaning as in that section.

The fundamental solution Λ constructed in Theorem 4 is the real part of an [a,b] pseudo-analytic function whose imaginary part is easily seen to be multiple-valued. Hence, if $\phi(z)$ is a single-valued solution of (2.2) defined for $0 < |z-z_0| < R$, or more generally, for $R_0 < |z-z_0| < R$, there exists a real constant C such that

---

[19] See [13]. This paper also contains the proofs of Lemma 7.

$\phi(z) - C\Lambda(z_0;z) = \text{Re}\,\{w(z)\}$, w being a single-valued $[a,b]$ pseudo-analytic function. Thus we obtain from the similarity principle the representation

$$(8.1) \qquad \phi(z) = C\Lambda(z_0;z) + \text{Re}\,\{e^{s(z)}f(z)\}$$

for every solution of (2.2) with an isolated singularity at z_0. Here $f(z)$ is an analytic function with a singularity at z_0, and $s(z)$ is of class H in a neighborhood of z_0. Also, to every f there is an s such that ϕ satisfies (2.2). The same representation holds for solutions possessing an isolated singularity at infinity, $f(z)$ being regular analytic for sufficiently large values of $|z|$. From (8.1) we obtain at once the classification of singularities (into removable, logarithmic, poles and essential).

On the other hand we have, by the expansion theorem for pseudo-analytic functions, the following

THEOREM 10. Every single-valued solution $\phi(z)$ of (2.2) defined for $0 \leq R_0 < |z-z_0| < R \leq +\infty$ admits in this domain the unique expansion

$$(8.2) \qquad \phi(z) = C\Lambda(z_0;z) + \sum_{-\infty}^{+\infty} A_n\Phi_n(z_0;z) + B_n\tilde{\Phi}_n(z_0;z).$$

Here Φ_n, $\tilde{\Phi}_n$ are the functions defined in §7. Theorem 10 contains an expansion theorem for a solution with a singularity at z_0 $(R_0 = 0)$ or at $z = \infty$ $(R = +\infty)$.

The behavior of a solution near a branch-point of finite order can be described similarly, but we do not discuss this here.

§9. Quasi-linear Elliptic Equations and Pseudo-Analytic Equations

We consider now a quasi-linear equation of the form

$$(9.1) \qquad A(p,q)r + 2B(p,q)s + C(p,q)t = 0,$$

for an unknown function $\phi(x,y)$, where $p = \phi_x$, $q = \phi_y$, $r = \phi_{xx}$, $s = \phi_{xy}$, $t = \phi_{yy}$. The real-valued coefficients A, B, C are assumed to be defined and of class H^2, and to satisfy the ellipticity condition

$$AC - B^2 = 1, \quad A > 0,$$

in a simply connected domain Ω (ellipticity domain) of the (p,q)-plane, containing the origin. A function $\phi(x,y)$ will be considered a solution of (9.1) only if $(\phi_x, \phi_y) \in \Omega$.

In view of Theorem 9 the linear system

$$\frac{\lambda_p}{A} = \frac{\lambda_q + \mu_p}{2B} = \frac{\mu_q}{C} \tag{9.2}$$

has in Ω a univalent solution $\lambda(p,q)$, $\mu(p,q)$. In view of the special form of (9.2), however, a stronger statement is true.

LEMMA 7.[19] System (9.2) possesses in Ω a solution (λ, μ) such that the mapping $(p,q) \longrightarrow (\lambda, \mu)$ is a homeomorphism of Ω onto a domain Ω^* and

$$\lambda = \mu = \lambda_q = 0, \qquad \lambda_p = 1 \quad \text{at} \quad p = q = 0 \tag{9.3a}$$

$$\lambda_p > 0 \tag{9.3b}$$

$$\lambda_p \mu_q - \lambda_q \mu_p > 0 \tag{9.3c}$$

$$p\lambda + q\mu > 0 \quad \text{for} \quad p^2 + q^2 > 0. \tag{9.3d}$$

The lemma implies that equation (9.1) can be written in the form of a "conservation law,"[20]

$$\lambda(\phi_x, \phi_y)_x + \mu(\phi_x, \phi_y)_y = 0. \tag{9.1a}$$

Hence, to every solution ϕ of (9.1) there exists a conjugate function ψ such that

$$\psi_x = -\mu(\phi_x, \phi_y), \qquad \psi_y = \lambda(\phi_x, \phi_y). \tag{9.4}$$

A simple computation shows that ψ satisfies the equation

$$A^*(\psi_x, \psi_y)\psi_{xx} + 2B^*(\psi_x, \psi_y)\psi_{xy} + C^*(\psi_x, \psi_y)\psi_{yy} \tag{9.5}$$

where $A^*[-\mu(p,q), \lambda(p,q)] = A(p,q)$ and B^*, C^* are defined similarly. The ellipticity domain of (9.5) is Ω^*. We call equations (9.1) and (9.5) conjugate.

---

[20] Cf. Loewner [23]. System (9.2) is dual, in the sense of Loewner, to equation (9.1) written in the form of a system for p, q.

LEMMA 8. If ϕ is a solution of (9.1) and ψ the conjugate function, then $\omega = \phi + i\psi$ is an interior function, that is of the form $\omega(z) = f[\chi(z)]$ where χ is a homeomorphism and f analytic.

PROOF. For a non-constant ϕ, $\phi_x^2 + \phi_y^2 = 0$ only at isolated points. In fact, considering $A(\phi_x, \phi_y)$, $B(\phi_x, \phi_y)$, $C(\phi_x, \phi_y)$ as given functions of (x,y) we can bring equation (9.1) into canonical form and apply Carleman's theorem. Next $J = \phi_x\psi_y - \phi_y\psi_x = p\lambda + q\mu > 0$ except at isolated points in view of (9.3d). This implies the assertion (cf. the proof of Theorem 7).

In order to express solutions of (9.1) by pseudo-analytic functions we consider the uniquely determined homeomorphism $u = u(p,q)$, $v = v(p,q)$ of Ω onto a domain $u^2 + v^2 < R^2 \leq +\infty$ which is conformal[21] with respect to the metric

$$ds^2 = A(p,q)dp^2 + 2B(p,q)dp\,dq + C(p,q)dq^2$$

and normalized by the requirement $u = v = u_q = 0$, $u_p = 1$ at $p = q = 0$. We call $w(p-iq) = u(p,q) - iv(p,q)$ the <u>distorted gradient</u>, and R the <u>conformal radius</u>.

Every function $\chi(p,q)$ may be considered as a function of w. Such a function will be said to <u>grow slowly</u> if either

$$(9.6)\qquad \chi = o(\log|w|), \qquad |w| \longrightarrow R = +\infty,$$

or

$$(9.7)\qquad \chi = O([R-|w|]^{\varepsilon-1}), \qquad |w| \longrightarrow R < +\infty, \qquad \varepsilon > 0,$$

In particular we set

$$(9.8)\qquad E(w) = A + C,$$

$$(9.9)\qquad F(w) = C, \qquad G(w) = i - B,$$

$$(9.10)\qquad \hat{F}(w) = \lambda F + \mu G, \qquad \hat{G}(w) = pG - qF,$$

$$(9.11)\qquad \Lambda(w) = |(G/F)_z/\mathrm{Im}\,\{G/F\}| \,.$$

We call E the <u>eccentricity</u> and Λ the <u>characteristic</u> of equation (9.1).

[21] The existence of the mapping (u,v) follows from Lichtenstein's theorem on the conformal mapping of non-analytic surfaces and from the general uniformization theorem.

It can be shown[22] that if E grows slowly, then $R = +\infty$ or $R < +\infty$ depending on whether Ω is or is not the whole plane.

Now let $\phi(x,y)$ be a fixed solution of (9.1) defined in a domain D, ψ its conjugate function and w its distorted gradient. We map the domain D onto a domain Δ in the ζ-plane conformally with respect to the metric

$$ds^2 = C\,dx^2 - 2B\,dx\,dy + A\,dy^2$$

and obtain a parametric representation of ϕ in the form

$$\text{(9.12)} \qquad \begin{cases} w = w(\zeta) \\ x + iy = z(\zeta) \\ \phi + i\psi = \omega(\zeta) \end{cases}$$

(The same procedure can be applied if ϕ is multiple-valued, but ϕ_x, ϕ_y are single-valued on a covering surface $\tilde{D}$ of D, provided $\tilde{D}$ is of genus zero.) It is easy to see that the functions z, ω are connected by the relation

$$\text{(9.12a)} \qquad (p\lambda+q\mu)(dx+idy) - (\lambda+i\mu)d\phi - i(p+iq)d\psi = 0.$$

THEOREM 11. In the parametric representation (9.12) the function $w(\zeta)$ is analytic, $z(\zeta)$ is (F,G) pseudo-analytic of the second kind, $\omega(\zeta)$ is $(\hat{F},\hat{G})$ pseudo-analytic of the second kind, (F,G) and $(\hat{F},\hat{G})$ being determined by (9.8), (9.10). Conversely, if the functions $w(\zeta)$, $z(\zeta)$, $\omega(\zeta)$ have these properties, $|w| < R$, and (9.12a) holds, then (9.12) is a parametric representation of a solution of (9.1).

The proof, largely computational, shall not be given here. Theorem 11 is a generalization of the Monge-Weierstrass representation of minimal surfaces by analytic functions.

EXAMPLE. Let $\rho(Q)$ be a positive function of class H^2 defined for $0 \leq Q < Q_{cr} \leq +\infty$ and satisfying the condition

$$\text{(9.13)} \qquad \frac{d[Q\rho(Q)]}{dQ} > 0.$$

[22] See [11]. This paper also contains the proofs of Theorems 11, 12, 13, 15, 16.

We interpret Q as the speed, ρ as the density of a potential gas flow. Condition (9.13) states that the flow is subsonic. The potential equation

(9.14) $$(\rho\phi_x)_x + (\rho\phi_y)_y = 0, \quad \rho = \rho(Q), \quad Q^2 = \phi_x^2 + \phi_y^2,$$

is of the form (9.1) and has the ellipticity domain $\phi_x^2 + \phi_y^2 < Q_{cr}^2$. The distorted gradient is easily computed to be

$$w = (\phi - i\phi_y)\exp \int_0^Q [X(Q)-1]\,\frac{dQ}{Q}, \quad X(Q)^2 = \frac{d[Q\rho(Q)]}{dQ}.$$

The stream-function of the flow satisfies the equation conjugate to (9.14)

$$(\psi_x/\rho)_x + (\psi_y/\rho)_y = 0,$$

with the ellipticity domain $\psi_x^2 + \psi_y^2 \leq Q_{cr}^{*2}$, where

(9.15) $$Q_{cr}^* = \lim_{Q \to Q_{cr}} [Q\rho(Q)].$$

Setting $\rho = [1-(\gamma-1)Q^2/2]^{1/(\gamma-1)}$, $\gamma = \text{const.} > 1$, we obtain the equation of an adiabatic gas flow. In this case $Q_{cr}^2 = 2/(\gamma+1)$, $R < +\infty$, $Q_{cr}^* < +\infty$, and the equation is of slowly growing eccentricity.

The equation of minimal surfaces

(9.16) $$(1+q^2)r - 2pq\,s + (1+p^2)t = 0$$

is obtained by setting $\rho = (1+Q^2)^{-1/2}$. For this equation $Q_{cr} = +\infty$, $R = 2$, $Q_{cr}^* = 1$, and $\Lambda \equiv 0$.

§10. Entire Solutions of Quasi-linear Equations

A well known theorem of S. Bernstein[23] states that all entire solutions of equation (9.16) are linear functions. The following proposition is a generalization of this result.

THEOREM 12.[22] If equation (9.1) has a finite conformal radius and a slowly growing characteristic, then all entire solutions of (9.1) are linear.

[23] Bernstein [5] obtained this result as a corollary of a geometrical theorem, cf. also E. Hopf [20], Mickle [24]. Other proofs were given by T. Rado [27], the author [7], and Heinz [19].

Necessary and sufficient conditions for the existence of non-linear entire solutions are known only in a relatively simple case.

THEOREM 13.[22] An equation (9.1) of slowly growing eccentricity possesses non-linear entire solutions if and only if Ω is the whole plane.

Both theorems assert the linearity of all entire solutions only under the assumption that $R < +\infty$. Simple examples show that the condition $R < \infty$ alone is insufficient.[24]

§11. Singularities of Solutions of Quasi-linear Equations

Sometime ago I proved that single-valued solutions of the equation (9.16) of minimal surfaces possess at finite points only removable isolated singularities.[25] In order that such a theorem hold for an equation of the form (9.14) it is necessary that

$$Q^*_{cr} < +\infty \tag{11.1}$$

(cf. (9.15)). In fact, if $Q^*_{cr} = +\infty$, then (9.14) has a solution of the form $\phi = \Phi(r)$, $r^2 = x^2 + y^2$, which becomes singular for $r = 0$. I conjectured that this condition is also sufficient.[26] R. Finn not only verified this but also obtained a more general theorem which can be restated as follows.

THEOREM 14. (Finn[27]) Assume that the ellipticity domain Ω of equation (9.1) is convex and the ellipticity domain Ω^* of equation (9.5) conjugate to (9.1) is bounded. Then if $\phi(z)$ is a single-valued solution of (9.1) defined for $0 < |z-z_0| < r$, $\phi(z_0)$ may be defined so that $\phi(z)$ becomes a regular solution of (9.1) for $|z-z_0| < r$.

---

[24] Sufficient conditions for Bernstein's theorem were also obtained by R. Finn (oral communication). Their relation to our conditions has not been clarified.

[25] Cf. [7], Theorem IX.

[26] [6], p. 163. There it is also conjectured that (11.1) implies Bernstein's theorem, but this is probably wrong.

[27] Finn [16].

Finn begins his proof by showing, by an elementary but intricate argument, that

$$\phi(z) = O(1), \quad z \longrightarrow z_o. \tag{11.2}$$

This also follows from Lemma 8. In fact, by (9.4) and the hypothesis, $\psi_x^2 + \psi_y^2 = O(1)$, so that $\psi(z)$ is single-valued and attains a limit ψ_0 as $z \longrightarrow z_0$. A homeomorphism of $0 < |z-z_0| < r$ onto a domain, which may be chosen as $0 \leq R_1 < |\zeta| < R_2$, takes $\phi + i\psi$ into an analytic function of ζ. Assume that $|\zeta| \longrightarrow R_1$ for $z \longrightarrow z_0$. Then $\psi \longrightarrow \psi_0$ for $|\zeta| \longrightarrow R_1$, and hence $\phi = O(1)$, either by the theorem on removable singularities (if $R_1 = 0$) or by the reflection principle (if $R_1 > 0$).

If Ω is the whole plane, the proof can be completed, following Finn, by choosing an r_0, $0 < r_0 < r$ and considering the solution $\tilde{\phi}(z)$ of the boundary value problem for (9.1) in the domain $|z-z_0| < r_0$ with boundary values $\phi(z)$. Set $\lambda = \lambda(\phi_x,\phi_y)$, $\tilde{\lambda} = \tilde{\lambda}(\tilde{\phi}_x,\tilde{\phi}_y)$, $\mu = \mu(\phi_x,\phi_y)$, $\tilde{\mu} = \tilde{\mu}(\tilde{\phi}_x,\tilde{\phi}_y)$. By (9.1a)

$$(\tilde{\lambda}-\lambda)(\tilde{\phi}_x-\phi_x) + (\tilde{\mu}-\mu)(\tilde{\phi}_y-\phi_y) = [(\tilde{\lambda}-\lambda)(\tilde{\phi}-\phi)]_x + [(\tilde{\mu}-\mu)(\tilde{\phi}-\phi)]_y.$$

Noting that the difference $\tilde{\phi} - \phi$ is bounded and vanishes for $|z-z_0| = r_0$, and applying Green's theorem we conclude that

$$\iint_{\eta < |z-z_0| < r_0 - \eta} [(\tilde{\lambda}-\lambda)(\tilde{\phi}_x-\phi_x)+(\tilde{\mu}-\mu)(\tilde{\phi}_y-\phi_y)]\,dx\,dy \longrightarrow 0 \tag{11.3}$$

$$\text{for} \quad \eta \downarrow 0.$$

Since at every point z the integrand in (11.3) is, in view of the mean-value theorem and of (9.2), a definite quadratic form in $(\tilde{\phi}_x-\phi_x)$, $(\tilde{\phi}_y-\phi_y)$, we conclude that $\phi = \tilde{\phi}$, q.e.d. For the case when Ω is not the whole plane, see Finn's paper.

We also have

> THEOREM 15.[22] Assume that the conformal radius of equation (9.1) is finite and the eccentricity grows slowly. Then if $\phi(z)$ is a solution of (9.1) defined for $0 < |z-z_0| < r$ and having single-valued derivatives ϕ_x, ϕ_y, then $\phi(z)$ is single-valued and the conclusion of Theorem 14 holds.

Simple examples show, however, that the conclusion of Theorem

14 is neither implied by nor implies the non-existence of non-linear entire solutions.

Singularities of solutions of (9.1) at $z = \infty$ must be considered separately.

THEOREM 16.[22] Assume that the conformal radius of equation (9.1) is finite and that its eccentricity grows slowly. Let $\phi(x,y)$ be a single-valued solution of (9.1) defined for all sufficiently large values of $x^2 + y^2$. Then ϕ_x and ϕ_y approach finite values as $x^2 + y^2 \longrightarrow \infty$.

The conclusion holds also if ϕ is not single-valued, but ϕ_x and ϕ_y are finitely-many-valued.

§12. A Non-linear Boundary Value Problem

Applications of pseudo-analytic functions to linear boundary value problems were mentioned in §5. Applications to boundary value problems for equations of the form (9.1) are also possible, but thus far only equations of the "gas-dynamical" form (9.15) have been considered.

Two-dimensional airfoil theory leads to the following problem. Let P be a simple closed curve possessing a Hölder-continuously turning tangent, except perhaps at one point z_T, at which there may be a protruding corner. A solution $\phi(z)$ defined in the domain exterior to P represents a flow past P if (i) ϕ_x, ϕ_y are single-valued and $\phi_x \longrightarrow Q_\infty > 0$, $\phi_y \longrightarrow 0$ as $z \longrightarrow \infty$, (ii) the normal derivative ϕ_n exists and vanishes on P, (iii) $\phi_x = \phi_y = 0$ at z_T (Kutta-Joukowski condition). We also require that (iv) ϕ_x, ϕ_y be Hölder-continuous on P. Problem $\Pi(Q_\infty)$ consists in finding a function satisfying these conditions for a given value of Q_∞ not exceeding the "critical speed" Q_{cr}.

THEOREM 17. There exists a positive number $\hat{Q} \leq Q_{cr}$ (depending on P) such that for $0 < Q_\infty < \hat{Q}$ Problem $\Pi(Q_\infty)$ has a solution ϕ which is unique (except for an additive constant). ϕ depends continuously on Q_∞, and as Q_∞ goes from 0 to $\hat{Q}$, $Q_{max} = \max|\phi_x + i\phi_y|$ takes on all values between 0 and Q_{cr}. (If P is symmetric with respect to the x-axis, Q_{max} is an increasing function of Q_∞.)

The first existence and uniqueness proof for a subsonic gas flow past a given (smooth) obstacle, with arbitrarily high subsonic values of the maximum speed, was given by Shiffman by variational methods.[28] Our proof (which will be published separately[29]) uses the theory of quasi-conformal and pseudo-analytic functions. The result is stronger than Shiffman's in two respects: the circulation is not prescribed but is shown to be uniquely determined by the Kutta-Joukowski condition; the uniqueness of the solution is established without assuming the solution to have a finite Dirichlet integral.

Recently P. W. Berg applied the same methods in proving the existence of a subsonic Helmholtz flow of a compressible fluid.[30]

---

[28] Shiffman [28]. Previously I obtained an existence theorem for the case of the equation of minimal surfaces [8]. For sufficiently small values of Q_∞, Problem $\Pi(Q_\infty)$ was solved by Frankl and Keldysh [17].

[29] Cf. [12].

[30] Cf. Berg [2].

BIBLIOGRAPHY

[1] AGMON, S. and BERS, L., "The expansion theorem for pseudo-analytic functions," Proceedings of the American Mathematical Society, 3 (1952), 757-764.

[2] BERG, P. W., "On the existence of Helmholtz flows of a compressible fluid," Thesis, New York University (1953).

[3] BERGMAN, S., "Linear operators in the theory of partial differential equations,"Transactions of the American Mathematical Society, 53 (1943), 130-155.

[4] BERGMAN, S., "Operator methods in the theory of compressible fluids," Proceedings of Symposia in Applied Mathematics, 1 (1949), 19-40.

[5] BERNSTEIN, S. N., "Sur un théorème de géométrie et ses applications aux équations aux dérivées partielles du type elliptique," Communications de la Sociéte Mathématique de Kharkov, 15 (1915-17), 38-45.

[6] BERS, L., "Singularities of minimal surfaces," Proceedings of the International Congress of Mathematicians, 2 (1950), 157-164.

[7] BERS, L., "Isolated singularities of minimal surfaces," Annals of Mathematics, 53 (1951), 364-386.

[8] BERS, L., "Boundary value problems for minimal surfaces with singularities at infinity," Transactions of the American Mathematical Society, 70 (1951), 465-491.

[9] BERS, L., "Partial differential equations and generalized analytic functions," Proceedings of the National Academy of Sciences, U. S. A., 36 (1950), 130-136. Second note, ibid. 37 (1951), 42-47.

[10] BERS, L., "Theory of pseudo-analytic functions," (Mimeographed lecture notes), New York University (1953).

[11] BERS, L., "Non-linear elliptic equations without non-linear entire solutions," (to appear).

[12] BERS, L., "Existence and uniqueness of a subsonic compressible flow past a given profile," Communications on Pure and Applied Mathematics, (to appear).

[13] BERS, L., "Univalent solutions of linear elliptic systems," Communications on Pure and Applied Mathematics, 6 4 (1953), 513-526.

[14] CARLEMAN, T., "Sur les systèmes linéaires aux dérivées partielles du premier ordre à deux variables," Comptes Rendus des séances de l'Académie des Sciences (Paris), 197 (1933), 471-474.

[15] DRESSEL, F. G. and GERGEN, J. J., "Mapping by p-regular functions," Duke Mathematical Journal, 18 (1951), 185-210.

[16] FINN, R., Thesis, Syracuse University (1952).

[17] FRANKL, F. I. and KELDYSH, M., "Die äussere Neumannsche Aufgabe für nichtlineare elliptische Differentialgleichungen mit Anwendung auf die Theorie der Flügel im kompressibilen Gas," Bulletin of the Academy of Sciences of the USSR. Mathematics. 12 (1943), 561-607.

[18] HADAMARD, J., Leçons sur la Propagation des Ondes et les Équations de l'Hydrodynamique, Hermann et Cie., Paris (1903).

[19] HEINZ, E., "Über die Lösungen der Minimalflächengleichung Göttingener Nachrichten (1952), 51-56.

[20] HOPF, E., "On S. Bernstein's theorem on surfaces $z(x,y)$ of non-positive curvature," Proceedings of the American Mathematical Society, 1 (1950), 80-85.

[21] JOHN, F., "The fundamental solution of linear elliptic differential equations with analytic coefficients," Communications on Pure and Applied Mathematics, 3 (1950), 273-304.

[22] JOHN, F., "Derivatives of solutions of linear elliptic partial differential equations," this Study.

[23] LOEWNER, C., "Conservation laws of certain systems of partial differential equations and associated mappings," this Study.

[24] MICKLE, E. J., "A remark on a theorem by Serge Bernstein," Proceedings of the American Mathematical Society, 1 (1950), 86-89.

[25] MIŠKIS, A., "Uniqueness of the solution of Cauchy's problem" Uspekhi Matematicheskikh Nauk, 3 (1948), 3-46.

[26] PFLUGER, A., "Quasikonforme Abbildungen und logarithmische Kapazität, Annales de l'Institut Fourier, 2 (1950), 69-80.

[27] RADÓ, T., "Zu einem Satze von S. Bernstein über die Minimalflächen im Grossen," Mathematiche Zeitschrift, 26 (1927), 559-565.

[28] SHIFFMAN, M., "On the existence of subsonic flows of a compressible fluid," Journal of Rational Mechanics and Analysis, 1 (1952), 605-652.

[29] VEKUA, I. N., New Methods for Solving Elliptic Equations, Moscow-Leningrad (1948).

[30] VEKUA, I. N., "Systems of partial differential equations of first order of elliptic type and boundary value problems with applications to the theory of shells," Matematiceskii Sbornik, 31, 73 (1952), 217-314.

VI. ON A GENERALIZATION OF QUASI-CONFORMAL MAPPINGS AND ITS APPLICATION TO ELLIPTIC PARTIAL DIFFERENTIAL EQUATIONS

L. Nirenberg

§1

Ever since the classical work of S. Bernstein [1], the question of the differentiability and analyticity of solutions of elliptic partial differential equations has received considerable attention. This note presents a result on the differentiability of solutions $z(x_1,\ldots,x_n)$ of the general nonlinear elliptic equation of second order

$$F(x_1,\ldots,x_n,z,z_1,\ldots,z_n,z_{11},\ldots,z_{nn}) = 0 \tag{1}$$

where

$$z_i = \frac{\partial z}{\partial x_i}, \quad z_{ij} = \frac{\partial^2 z}{\partial x_i \partial x_j}, \quad i,j = 1,\ldots,n.$$

Only a brief description of its proof and related results are presented here; the complete proofs will appear elsewhere.

The strongest results for such equations are due to G. Giraud [2] and E. Hopf [3]. Assuming that z is twice continuously differentiable and that the second derivatives of z are Hölder continuous[1] they proved (a) if F has Hölder continuous partial derivatives up to order m with respect to all of its arguments then z possesses Hölder continuous partial derivatives up to order $m+2$, (b) if F is analytic in its arguments then z is an analytic function.

Our main result asserts that their assumption of the Hölder continuity of the second derivatives of a solution z of (1) is superfluous

[1] A function f defined in a set is said to satisfy a Hölder condition (or inequality) in that set, with positive constants $\alpha < 1$ and H called the exponent and coefficient respectively of the condition, if for any two points P, Q of the set the inequality $|f(P)-f(Q)| \le H\overline{PQ}^{\alpha}$ holds, where $\overline{PQ}$ denotes the distance between P and Q. A function defined in a domain D is said to be Hölder continuous in D if it satisfies a Hölder condition in every closed subdomain of D.

because it may be derived:

THEOREM 1. Let $z(x_1,\dots,x_n)$ be a twice continuously differentiable solution of the elliptic equation (1) in a domain D. Assume that F possesses continuous first derivatives with respect to all arguments. Then the second derivatives of z are Hölder continuous in D.[2]

For $n = 2$ this result was established in an earlier paper[3] [5] with the aid of a lemma concerning the uniform Hölder continuity of a class of mappings (which includes quasi-conformal mappings) of the plane into itself. For $n > 2$ Theorem 1 is deduced from a similar theorem about the uniform Hölder continuity of a class of such mappings in n-space which is a generalization of quasi-conformal mappings in the plane. These are defined by functions $p_i(x_1,\dots,x_n)$, $i = 1,\dots,n$, satisfying an inequality of the form

$$(2) \qquad \sum_{i,j=1}^{n} p_{ij}^2 \leq k \sum_{i,j=1}^{n} (p_{ij}p_{ji}-p_{ii}p_{jj}) + k_1$$

where

$$p_{ij} = \frac{\partial p_i}{\partial x_j}$$

Here k and k_1 are non-negative constants with the restriction

$$(3) \qquad k < \frac{n-1}{n-2}$$

Consider mappings with $k_1 = 0$; for $n = 2$, condition (3) is no restriction and such mappings are quasi-conformal. For $n > 2$ this is no longer the case, but such mappings do possess some of the properties of quasi-conformal mappings in the plane. It is easily seen (with $k_1 = 0$) that if $k < 1$ the mapping functions are constant, while if $k = 1$ they

---

[2] This result for general second order elliptic systems has been derived by a different method in the paper by C. B. Morrey in this volume.

[3] For $n = 2$ the constants of the Hölder inequality for any closed subdomain depend (essentially) only on bounds for the first and second derivatives of the solution z; and the result may therefore be used to derive a priori estimates for all higher order derivatives of the solution. For $n > 2$, however, the constants of the Hölder inequality depend, in addition, on the modulus of continuity of the second derivatives of z, and the result is therefore not as useful.

are the first derivatives of a harmonic function (the analogue of the Cauchy-Riemann equations for $n = 2$).

The statement of the uniform Hölder continuity of such mappings (with fixed k and k_1) is

LEMMA 1. Let the functions $p_1, \ldots, p_n$ be once continuously differentiable and satisfy (2) in a domain D, and assume $|p_i| \leq$ a constant K $(i = 1, \ldots, n)$. In any closed subdomain B of D the functions p_i satisfy a Hölder condition with constants depending only on k, k_1, K and the distance from B to the boundary of D.

§2

In proving Theorem 1 we first reduce its proof to that of a result concerning linear elliptic differential equations (Theorem 2). This is accomplished by the following device: It is easily seen that the difference quotient $z^h = \frac{1}{h}[z(x_1+h, x_2, \ldots, x_n) - z(x_1, x_2, \ldots, x_n)]$ of a solution z of (1), for any $h > 0$, satisfies a linear elliptic equation of the form

$$\sum_{i,j=1}^{n} a_{ij}\, z^h_{ij} = f$$

where the coefficients f and a_{ij} $(i, j = 1, \ldots, n)$ are known functions (involving of course the solution z) which depend on h. These coefficients satisfy certain inequalities in D:

$$|f| \leq K_1 \tag{4}$$

$$M \sum_i \xi_i^2 \geq \sum_{i,j} a_{ij} \xi_i \xi_j \geq m \sum_i \xi_i^2 \qquad \text{for all real } \xi_i \tag{5}$$

with positive constants K_1, M and m independent of h. Furthermore, the coefficients a_{ij} possess a uniform modulus of continuity independent of h.

Thus the Hölder continuity of the derivatives z_{1i} $(i = 1, \ldots, n)$ of z (and, by a similar argument, all other second derivatives of z) may be derived as a consequence of the following theorem, by letting $h \longrightarrow 0$.

THEOREM 2. Let u be a twice continuously differentiable solution of an elliptic equation

(6) $$\sum_{i,j=1}^{n} a_{ij} u_{ij} = f$$

in a domain D. Assume that the coefficients satisfy (4) and (5) with certain positive constants K_1, M and m, and that the coefficients a_{ij} are continuous in the closure of D. If the first derivatives of u are bounded in absolute value by a constant K then they satisfy a Hölder condition in any closed subdomain B of D with constants depending only on K_1, M, m, K, the modulus of continuity of the coefficients a_{ij}, and the distance from B to the boundary of D.

For $n = 2$ this result is due to C. B. Morrey [4] and is also proved in [5]. In that case the constants of the Hölder condition are independent of the continuity of the coefficients a_{ij}.

The proof of Theorem 2 follows from Lemma 1 and is based on the fact that, if the independent variables are transformed so that the differential operator in (6) is the Laplacean at a point, then the first derivatives of a solution u of (6) define a mapping of the kind considered in Lemma 1 with constants k and k_1 depending only on K_1, M, m and K — at least in a region about the point whose size depends on the modulus of continuity of the a_{ij}. This may be seen as follows. In a neighborhood of the point the differential equation takes the form

$$\sum_{i} u_{ii} = \sum_{i,j} (\delta_{ij}-a_{ij})u_{ij} + f$$

where δ_{ij} is the Kronecker delta. If now both sides are squared, and consideration is restricted to a region about the point in which the inequality

(7) $$\text{l.u.b.} \sum_{i,j} (\delta_{ij}-a_{ij})^2 < \frac{1}{n-1}$$

holds, then it is easily seen that the first derivatives u_i of u satisfy in this region an inequality of the form (2) with constants

$$k < \frac{n-1}{n-2}$$

and k_1 which may be determined. Application of Lemma 1 yields the Hölder inequality for the u_i in a subregion, and application of the argument over a number of regions yields the proof of Theorem 2.

§3

For $n = 2$ a proof of Lemma 1 was given by the author in [5], §3, using modifications of techniques of Morrey; the proof in the more general case is similar. The essence of the technique is to derive estimates for certain integrals involving the derivatives of the p_i, of the form

$$\int r^{2-n-\alpha} \sum_{i,j} p_{ij}^2 \, dv \leq H$$

where the integration extends over a sphere with dv as element of volume; r represents the distance from the center of the sphere. Here $\alpha < 1$ and H are fixed positive constants depending on the constants in Lemma 1. These estimates then yield a Hölder inequality for the functions p_i with exponent $\frac{\alpha}{2}$ and fixed coefficient.

These integral estimates may also be used to derive a Liouville-type theorem for mapping functions satisfying (2) (and (3)) with $k_1 = 0$:

LEMMA 2. Let $p_1, \ldots, p_n$ be such a mapping defined in the whole space and assume that $n-1$ of the mapping functions are bounded. Then the mapping is identically constant.

This lemma may be applied directly to the homogeneous equation corresponding to (6) and yields

THEOREM 3. Let u be a solution of the elliptic equation

$$\sum_{i,j} a_{ij} u_{ij} = 0$$

over the whole space, and assume that (7)

$$\text{l.u.b.} \sum_{i,j} (a_{ij} - \delta_{ij})^2 < \frac{1}{n-1}$$

is satisfied.[4] If $n-1$ of the first order derivatives of u are bounded then u is a linear function.

REMARKS: The theorems and lemmas are not stated here in their sharpest form. For instance the conditions in Lemma 1 and Theorem 2 on the continuity of the derivatives of the p_i and u can be relaxed. In addition, in Lemma 2 and Theorem 3 the condition of boundedness of the p_i and u_i can be weakened.

It should be noted that the restriction (3) cannot be dropped. For n even, it is easy to construct explicit mappings with

$$k = \frac{n-1}{n-2}$$

and satisfying, otherwise, the conditions of Lemmas 1 and 2 for which the conclusions of these lemmas are not valid.

---

[4] Probably much too restrictive. For $n = 2$ it suffices that the equation have uniform ellipticity over the plane, i.e., that the coefficients satisfy an inequality of the form (5) over the whole plane with fixed positive constants M and m.

BIBLIOGRAPHY

[1] BERNSTEIN, S., "Sur la nature analytique des solutions des equations aux dérivées partielles du second ordre," Mathematische Annalen 59 (1904), 20-76.

[2] GIRAUD, G., "Équations de type elliptique," Annales Scientifiques de l'École Supérieure 47 Paris (1930), 197-266 and references quoted here to Giraud's previous work.

[3] HOPF, E., "Über den funktionalen, insbesondere den analytischen Character der Lösungen elliptischer Differentialgleichungen zweiter Ordnung," Mathematische Zeitschrift 34 No. 2 (1931), 191-233.

[4] MORREY, C. B., "On the solutions of quasilinear elliptic partial differential equations," Transactions of the American Mathematical Society 43 (1938), 126-166.

[5] NIRENBERG, L., "On nonlinear elliptic partial differential equations and Hölder continuity," Communications on Pure and Applied Mathematics 6 No. 1 (1953).

VII. SECOND ORDER ELLIPTIC SYSTEMS OF DIFFERENTIAL EQUATIONS

C. B. Morrey, Jr.

INTRODUCTION

In this paper, we study the existence and regularity properties of the solutions of linear elliptic systems of partial differential equations in N dependent and ν independent variables of the form

$$\text{(A)} \qquad \int_{R^*} (a_{ij}^{\alpha\beta} u^j_{x^\beta} + b_{ij}^{\alpha} u^j + e_i)\, dx'_\alpha = \int_R (c_{ij}^{\alpha} u^j_{x^\alpha} + d_{ij} u^j + f_i)\, dx$$

$$i = 1, \ldots, N, \qquad \alpha, \beta = 1, \ldots, \nu, \qquad dx'_\alpha = (\zeta_\alpha \cdot n)\, dS$$

ζ_α being the unit vector in the x^α direction and n the exterior normal. The minimum assumptions on the tensors $a(x)$ to $f(x)$ are that $a(x)$ be continuous, $b(x)$, $c(x)$, and $d(x)$ be bounded and measurable, and the $e(x)$ and $f(x)$ be in L_2. The solutions u are required to be of class $\mathfrak{P}_2$ on domains interior to G and to satisfy (A) on "almost all cells" R in G (see [5]). The ellipticity condition is

$$\text{(B)} \quad \text{determinant of } N \times N \text{ matrix } \|a_{ij}^{\alpha\beta}(x)\lambda_\alpha \lambda_\beta\| \neq 0, \quad \lambda \neq 0, \quad x \in G,$$

$$\lambda = (\lambda_1, \ldots, \lambda_\nu), \qquad x = (x^1, \ldots, x^\nu)$$

In (A) and (B), we have used the following tensor summation convention which will be used throughout the paper: repeated Greek indices are summed from 1 to ν and Latin indices from 1 to N.

We have selected the Haar or integrated form (A) in order to obtain regularity properties of the solutions under very weak assumptions concerning the coefficients. We have also considered systems of the form

$$\text{(C)} \qquad a_{ij}^{\alpha\beta} u^j_{x^\alpha x^\beta} = c_{ij}^{\alpha} u^j_{x^\alpha} + d_{ij} u^j + f_i$$

with the same assumptions on the coefficients; here we require that the solutions u and their derivatives be of class $\mathfrak{P}_2$ on interior domains and that u satisfy (C) almost everywhere. The regularity results for the systems (A) and (C) which hold on interior domains are proved in §4 under various hypotheses about $a(x)$ to $f(x)$, using only the general ellipticity condition (B). The results for the system (C) include results analogous to those obtained by Schauder (in [13] and [14]), Giraud, and others for the case $N = 1$. In §4, it is shown that if the tensors $a(x)$, $b(x)$, and $e(x)$ are of class C' with μ-Hölder-continuous first derivatives and if $c(x)$, $d(x)$, and $f(x)$ are μ-Hölder-continuous, then the solutions u of (A) are of class C'' with μ-Hölder continuous second derivatives on interior domains and hence satisfy

$$\text{(D)} \qquad \frac{\partial}{\partial x^\alpha}\left(a^{\alpha\beta}_{ij} u^j_{x^\beta} + b^\alpha_{ij} u^j + e^\alpha_i\right) = c^\alpha_{ij} u^j_{x^\alpha} + d_{ij} u^j + f_i$$

The higher order regularity properties for (A) on the interior then follow from those for the system (C).

The results allowing more general coefficients are more in the tradition of the work of E. Hopf [6] and the writer ([10] and [12]). The results of F. John in [7], adapted in §1 to the general "pure" second order elliptic system with constant coefficients, are fundamental to the discussion of the general elliptic system.

In order to obtain the existence theorems given in §8 and the regularity properties on the boundary given in §7, the writer has found it necessary to restrict himself to systems (A) satisfying the <u>strong ellipticity condition</u>

$$\text{(E)} \qquad a^{\alpha\beta}_{ij}(x)\lambda_\alpha\lambda_\beta\xi^i\xi^j > 0, \quad x \in \bar{G}, \quad \lambda \neq 0, \quad \xi \neq 0$$

The method of this paper in obtaining the existence and boundary value results requires that the solution of the Dirichlet problem for the pure second order system with constant coefficients be always possible and unique for arbitrary bounded domains. This has been shown by simple examples to be false in general for the general elliptic system (see [2]). A simple example is the following:

$$u^1_{x^1x^1} - u^1_{x^2x^2} - 2u^2_{x^1x^2} = 0$$

$$2u^1_{x^1x^2} + u^2_{x^1x^1} - u^2_{x^2x^2} = 0$$

which has the solutions $u^i = [1 - (x^1)^2 - (x^2)^2] \cdot H^i$, $i = 1, 2$, H^1 and H^2 being any conjugate harmonic functions, all of which solutions vanish on the boundary of the unit sphere; this system is elliptic, the determinant in (B) being $2|\lambda|^4$.

The existence theory given in §8 is very similar to the theories given by numerous authors for various classes of equations culminating with those of F. E. Browder ([3] and [4]) for the single equations of order $2m$ and M. I. Visik [17] for strongly elliptic systems of order $2m$. The main new feature in the existence theory of §8 consists in the weak assumptions made concerning the coefficients.

In §9, the results and methods of §§1 - 8 are applied to obtain further regularity results for the solutions, assumed to be of class C'', of general second order non-linear elliptic systems, and for the solutions, assumed to be of class C', of general regular variational problems.

If G is an open set, G^* denotes its boundary and $\overline{G}$ its closure; if $P \in G$, δ_P denotes the distance of P from G^*. We shall be dealing with various vector and tensor functions, frequently denoted by single letters, such as

$$u(x) = [u^1(x), \ldots, u^N(x)], \quad \nabla u(x) = \left\{u^i_{x^\alpha}(x)\right\}, \quad \nabla^2 u(x) = \left\{u^i_{x^\alpha x^\beta}(x)\right\}$$

$$a(x) = a^{\alpha\beta}_{ij}(x), \quad b(x) = b^{\alpha}_{ij}(x), \text{ etc.}, \quad (X)^n = (x^\alpha)^n = \left\{x^{\alpha_1} \cdots x^{\alpha_n}\right\}, \text{ etc.}$$

$\nabla^2 u$ does not denote the Laplacian. If ϕ is any vector or tensor, $|\phi|^2$ denotes the sum of the squares of all its components. We shall some times abbreviate $u^i_{x^\alpha}$ to u^i_α and

$$a^{\alpha\beta}_{ij} u^j_{x^\alpha x^\beta} \text{ to } a \cdot \nabla^2 u, \quad b^\alpha_{ij} u^j \text{ to } b \cdot u, \text{ etc.}$$

Functions of class $\mathfrak{P}_2$ have been discussed at length in [5] and [11]. If u (i.e., its components) are of class $\mathfrak{P}_2$, we define

$$D_2(u, G) = \int_G |\nabla u|^2 \, dx$$

Similar notations are the following

$$L_2(u, G) = \int_G |u|^2 \, dx, \quad l(u, G) = d_0(u, G) = [L_2(u, G)]^{1/2}$$

$$d(u, G) = d_1(u, G) = [D_2(u, G)]^{1/2}, \quad d_k(u, G) = \left[\int_G |\nabla^k u|^2 \, dx\right]^{1/2}$$

The sphere with center P and radius a is denoted by $C(P, a)$. Finally u will be said to be of class $C^{(n)}_\mu$, $0 < \mu < 1$, $n \geq 0$, on $\overline{G}$ if u is of class $C^{(n)}$ on $\overline{G}$ (i.e., derivatives tend to limits on G^*) and $\nabla^n u$ satisfies a uniform Hölder condition with exponent μ on $\overline{G}$; u is of class $C^{(n)}_\mu$ on an <u>open</u> set G if and only if it is of class $C^{(n)}$ on each closed $\overline{D} \subset G$.

We define γ_ν and Γ_ν as the measure and surface-measure of the unit sphere in ν dimensions. We have occasion to separate out the dependence of a function $u(x)$ on the single variable x^ν; when this is desired we write $u(x) = u(x^\nu, x'_\nu)$, x'_ν denoting the other variables. We use several sequences of constants $C_1, C_2, \ldots, K_1, K_2, \ldots, Q_1, Q_2, \ldots$; of these, the C_n do not depend on the differential equations under consideration; constants Z_1, Z_2, etc., which occur in the proofs of theorems mean different things in different theorems (the same within the proof of one theorem).

§1. THE ELEMENTARY MATRIX; $\bar{a}$-HARMONIC FUNCTIONS

In this section, we consider solutions of the general elliptic system

$$(1.1) \qquad \bar{a}^{\alpha\beta}_{ij} u^j_{x^\alpha x^\beta} = 0 \quad (\bar{a} \text{ constant})$$

such solutions u are called $\bar{a}$-harmonic.

It is easy to apply the method of F. John [7] to determine an <u>elementary matrix</u> for the system (1.1). This is done as follows: Define the functions $A^{kj}(\lambda)$ so that

$$(1.2) \qquad (\bar{a}^{\alpha\beta}_{ij} \lambda_\alpha \lambda_\beta) A^{kj}(\lambda) = \delta^k_i = (\bar{a}^{\alpha\beta}_{ij}\lambda_\alpha\lambda_\beta) \cdot A^{jk}(\lambda), \quad \lambda \neq 0$$

the existence of these functions is guaranteed by the ellipticity condition (B). We then define $G(\sigma)$ so that

$$G(\sigma) = G(-\sigma), \quad G(0) = G'(0) = 0, \quad G''(\sigma) = \begin{cases} \log|\sigma|, & \nu \text{ even} \\ |\sigma|, & \nu \text{ odd and} \geq 3 \end{cases}$$

and we define

$$V^{kj}(y) = \int_{|\lambda|=1} G(y \cdot \lambda) A^{kj}(\lambda) d\textstyle\sum_\lambda = V^{kj}(-y)$$

$d\sum_\lambda$ being the element of hyper-area on $|\lambda| = 1$. Finally, we define

$$(1.3) \qquad \Gamma^{kj}(y) = \begin{cases} E_\nu \Delta^{\nu/2} V^{kj}(y), & \nu \text{ even} \\ F_\nu \Delta^{(\nu+1)/2} V^{kj}(y), & \nu \text{ odd} \end{cases}$$

Δ being the Laplacian operator, E_ν and F_ν being proper constants. The method of John can be followed through to show that the $\Gamma^{kj}(y)$ are analytic for all $y \neq 0$ and constitute an elementary matrix in the sense that

$$u^i(x) = \int_{G^*}\left[u^j(\xi)\bar{a}^{\beta\alpha}_{kj}\Gamma^{ik}_{\beta}(\xi - x) - \Gamma^{ik}(\xi - x)\bar{a}^{\alpha\beta}_{kj}u^j_{\beta}(\xi)\right] d\xi'_{\alpha}$$

(1.4)

$$+ \int_G \Gamma^{ik}(\xi - x)\bar{a}^{\alpha\beta}_{kj}u^j_{\alpha\beta}(\xi)\, d\xi$$

$$x \in G, \quad d\xi'_{\alpha} = \cos(\alpha, n)\, dS$$

whenever u is of class C'' on the closed region $\bar{G}$ with smooth boundary G^*. It is easy to see that

$$\bar{a}^{\alpha\beta}_{ji}\Gamma^{kj}_{\alpha\beta}(y) = \bar{a}^{\alpha\beta}_{ij}\Gamma^{jk}_{\alpha\beta}(y) = 0, \quad \Gamma^{ij}(-y) = \Gamma^{ij}(y), \quad y \neq 0$$

(1.5)

$$\int_{G^*}\bar{a}^{\alpha\beta}_{ji}\Gamma^{kj}_{\beta}(y)\, dy'_{\beta} = \delta^k_i, \quad y \in G, \ G \text{ smooth}$$

and that the Γ^{ij} are positively homogeneous of degree $2 - \nu$ if $\nu > 2$, or are of the form

$$\Gamma^{ij}(y) = c^{ij}\log|y| + \tilde{\Gamma}^{ij}(y)$$

if $\nu = 2$, the c^{ij} being constants and the $\tilde{\Gamma}^{ij}$ being positively homogeneous of degree 0.

We are interested in keeping track of bounds for our solutions. The ellipticity condition (B) guarantees the existence of a number M such that

(1.6) $\quad |a(x)| \leqq M \cdot (\nu N)^{1/2}, \quad \|\tau(x, \lambda)\| \leqq M, \quad x \in \bar{G}, \quad |\lambda| = 1$

where $\tau(x, \lambda)$ is the linear transformation from E_N to E_N given by

$$\tau(x, \lambda) : \eta^i = A^{ij}(x, \lambda)\xi^j$$

the $A^{ij}(x, \lambda)$ being defined by (1.2) in terms of $a(x)$. For Laplace's equation $(\bar{a}^{\alpha\beta}_{ij} = \delta^{\alpha\beta} \cdot \delta_{ij})$, we may take $M = 1$. It follows easily that there are constants K_0 (and K_0' if $\nu = 2$), K_1, K_2, and K_3, depending only on ν, N, and M such that

$$|\nabla\Gamma(y)| \leqq K_1|y|^{1-\nu}, \quad |\nabla^2\Gamma(y)| \leqq K_2|y|^{-\nu}, \quad |\nabla^3\Gamma(y)| \leqq K_3|y|^{-1-\nu}$$

$$(1.7) \qquad |\Gamma(y)| \leqq \begin{cases} K_0 + K_0'\,|\log|y||, & \nu = 2 \\ K_0|y|^{2-\nu}, & \nu > 2 \end{cases}$$

If u is $\bar{a}$-harmonic on a sphere $C(P, a)$, we obtain a representation of $u(x)$ for $x \in C(P,c)$, $c < a$, of the form

$$\begin{aligned}(a^2-c^2)(a^2-b^2)(b^2-c^2)u^i(x) &= 4(b^2-c^2)\int_{C_a-C_b}(a^2-|\xi|^2)[2\xi^{\alpha}\bar{a}_{kj}^{\alpha\beta}\Gamma_{\beta}^{ik}(\xi-x) \\ &\quad + \bar{a}_{kj}^{\alpha\alpha}\Gamma^{ik}(\xi-x)]u^j(\xi)\,d\xi \\ (1.8) \qquad &+ 4(a^2-b^2)\int_{C_b-C_c}(|\xi|^2-c^2)[2\xi^{\alpha}\bar{a}_{kj}^{\alpha\beta}\Gamma_{\beta}^{ij}(\xi-x) + \bar{a}_{kj}^{\alpha\alpha}\Gamma^{ik}(\xi-x)]u^j(\xi)\,d\xi \\ &- 8(b^2-c^2)\int_{C_a-C_b}\bar{a}_{kj}^{\alpha\beta}\xi^{\alpha}\xi^{\beta}\Gamma^{ik}u^j\,d\xi + 8(a^2-b^2)\int_{C_b-C_c}\bar{a}_{kj}^{\alpha\beta}\xi^{\alpha}\xi^{\beta}\Gamma^{ik}u^j\,d\xi\end{aligned}$$

$$b = (a+c)/2$$

This is done by taking (1.4) with $G = C(P, r) = C_r$, $c < r < b$, $x \in C_c$, multiplying by r, integrating with respect to r from r_1 to r_2, integrating by parts to get rid of the derivatives of u, multiplying by $16r_1r_2$ and integrating with respect to r_1 and r_2 over $[c, b]$ and $[b, a]$, respectively.

The representation (1.8) allows us to show that u is analytic in $C(P, a)$, that a weak solution u in $\mathfrak{P}_2$ of

$$(1.9) \qquad \int_{R^*}\bar{a}_{ij}^{\alpha\beta}u_{\beta}^j\,dx_{\alpha}' = 0$$

is $\bar{a}$-harmonic, and that the derivatives of u are $\bar{a}$-harmonic. We can also show that there are constants $K_4, K_5, \ldots, K_9$ depending only on ν, N, and M, such that

$$\text{(i)}\quad |\nabla^n u(P)| \leqq K_4 a^{-\rho} d_n[u, C(P, a)]$$

$$\text{(ii)}\quad |\nabla^{n+1} u(P)| \leqq K_5 a^{-\rho-1} d_n[u, C_a]$$

$$\text{(iii)}\quad |\nabla^{n+2} u(P)| \leqq K_6 a^{-\rho-2} d_n[u, C_a]$$

$$\text{(1.10)}\qquad \text{(iv)}\quad d_n(u, C_r) \leqq K_7 d_n(u, C_a) \cdot (r/a)^\rho$$

$$\text{(v)}\quad d_{n+1}(u, C_{a/2}) \leqq K_8 a^{-1} d_n(u, C_a)$$

$$\text{(vi)}\quad d_{n+2}(u, C_{a/2}) \leqq K_9 a^{-2} d_n(u, C_a)$$

$$(\rho = \nu/2)$$

where $K_7 = 1$ for Laplace's equation. Using (1.10), we see that a function which is $\bar{a}$-harmonic with finite Dirichlet integral over the whole of E_ν is a constant.

§2. SOLUTION OF A NON-HOMOGENEOUS SYSTEM OVER E_ν

In this section, we discuss certain solutions of

$$\text{(2.1)}\qquad \int_{R*} (\bar{a}^{\alpha\beta}_{jk} u^k_{x^\beta} + e^\alpha_j)\, dx'_\alpha = 0$$

in which the $\bar{a}$ are constants satisfying (1.6) and e is in L_2 on E_ν.

THEOREM 2.1. There is a solution u of (2.1) which is of class $\mathfrak{P}_2$ on any bounded domain with $d_1(u, E_\nu) < \infty$ and any two such solutions differ by a constant (and a null function). If $C(P, a)$ is any sphere, one such solution is given (almost everywhere) by

$$\text{(2.2)}\qquad u^j(x) = \int_{C(P,a)} \Gamma^{jk}_\alpha(\xi - x) e^\alpha_k(\xi)\, d\xi + \int_{E-C(P,a)} [\Gamma^{jk}_\alpha(\xi - x) - \Gamma^{jk}_\alpha(\xi - x_P)] e^\alpha_k(\xi)\, d\xi$$

Finally, if u is any such solution

$$\text{(2.3)}\qquad d(u, E_\nu) \leqq M l(e, E_\nu)$$

PROOF: It is clear that any two such solutions differ by a constant since the difference would satisfy (1.9) and hence be $\bar{a}$-harmonic.

If e is of class C'' on E_ν and vanishes outside some sphere, the classical procedure for differentiating (2.2) shows that u, defined by (2.2), is of class C'' everywhere and satisfies

$$\bar{a}^{\alpha\beta}_{jk} u^k_{\alpha\beta}(x) = - e^{\alpha}_{jk\alpha}(x) \tag{2.4}$$

and hence satisfies (2.1). Applying Schwarz's inequality, etc., to (2.2), we see that

$$l[u, C(P, a)] \leqq Z_1 \cdot a \cdot l(e, E_\nu), \qquad Z_1 = Z_1(\nu, N, M) \tag{2.5}$$

We also see that ∇u and $\nabla^2 u$ are in L_2 over E_ν. If we take Fourier transforms (with the factor $(2\pi)^{-\nu/2}$), and denote the transform of $e^{\alpha}_j(x)$ by $E^{\alpha}_j(y)$ and those of $u^j_{x^\alpha}$ and $u^j_{x^\alpha x^\beta}$ by $U^j_\alpha(y)$ and $U^j_{\alpha\beta}(y)$, we see that there is a function $U(y)$ with $|y|\,U(y)$ (and in this case, also with $|y|^2 U(y)$) in L_2 such that

$$\begin{gathered} U^j_\alpha(y) = - i y^\alpha U^j(y), \qquad U^j_{\alpha\beta}(y) = - y^\alpha y^\beta U^j(y) \\ d(u, E_\nu) = \left[\int_{-\infty}^{\infty} |y|^2 \cdot |U(y)|^2 \, dy\right]^{1/2}, \qquad l(e, E_\nu) = l(E, E_\nu) \end{gathered} \tag{2.6}$$

thus, from (2.4), we have

$$(\bar{a}^{\alpha\beta}_{jk} y^\alpha y^\beta) U^k(y) = i y^\alpha E^{\alpha}_j(y) \tag{2.7}$$

so that we can deduce (2.3) from (2.6), (2.7), (1.6) and (1.2). The result for the general e in L_2 on E_ν follows from (2.5) and (2.3) by a straightforward limiting process.

We now work toward the important Dirichlet growth Theorem 2.2. The following lemma is also useful later.

LEMMA 2.1. Suppose $\phi(s)$ is a non-decreasing function on $[0, 1]$ with $\phi(0) = 0$, $\phi(1) \leqq L_1$, and suppose

$$\phi(s) \leqq t^\lambda \phi(s/t) + L_2 (s/t)^\rho \phi(t) + L_3 L_1 s^\rho t^{\lambda-\rho} \tag{2.8}$$

$$0 < s \leqq t \leqq 1$$

$$0 \leqq \lambda < \rho = \nu/2$$

Then there is a number $S(L_2, L_3, \nu, \lambda)$ such that

$$\phi(s) \leqq L_1 S s^{\lambda}, \quad 0 \leqq s \leqq 1, \quad S \geqq 1$$

PROOF. Choose $0 < \sigma < 1$, and suppose

$$\phi(s) \leqq L_1 S_0 s^{\lambda}, \quad \sigma \leqq s \leqq 1, \quad S_0 \geqq 1$$

Applying (2.8), we see, since $S_0 \geqq 1$, that

$$\phi(s) \leqq L_1 S_0 s^{\lambda} \, [1 + L_4 (s/t)^{\rho-\lambda}], \; \sigma^2 \leqq s \leqq \sigma, \; s \leqq t = \sigma^{-1} s \leqq 1, \; L_4 = L_2 + L_3$$

Hence

$$\phi(s) \leqq L_1 S_1 s^{\lambda}, \; S_1 = S_0(1 + L_4 \tau), \; \tau = \sigma^{\rho-\lambda}, \; \sigma^2 \leqq s \leqq 1 \tag{2.9}$$

since $S_1 \geqq S_0$. Repeating, we obtain

$$\phi(s) \leqq L_1 S_n s^{\lambda}, \quad S_n = S_0 \prod_{p=0}^{n-1} (1 + L_4 \tau^{2^p}), \quad \sigma^{2^n} \leqq s \leqq 1 \tag{2.10}$$

Since $\tau < 1$, we may let $n \longrightarrow \infty$ in (2.10). Since we may obviously choose $S_0 = \sigma^{-\lambda}$, we may take S as the minimum of

$$\sigma^{-\lambda} \prod_{p=0}^{\infty} (1 + L_4 \tau^{2^p}), \quad 0 < \sigma < 1$$

THEOREM 2.2. Suppose e satisfies

$$l[e, C(P, r)] \leqq \begin{cases} L(r/a)^{\lambda}, & 0 \leqq r \leqq a, \quad 0 \leqq \lambda < \rho = \nu/2 \\ L \quad , & r \geqq a \end{cases} \tag{2.11}$$

at some point P. Then, if u is given by (2.2), we have

$$d[u, C(P, r)] \leqq K_{10}(\lambda, \nu, N, M) \cdot L \cdot (r/a)^{\lambda}, \; 0 \leqq r \leqq a$$

$$K_{10} = MS(K_7, K_7, \nu, \lambda) \geqq M$$

PROOF. We define $\phi(s)$ on $[0, 1]$ as the least upper bound of $L^{-1} d[u, C(P, as)]$ for all P, a, and e satisfying (2.11) at P, u

being given by (2.2). Clearly ϕ is non-decreasing with $\phi(0) = 0$; $\phi(1) \leqq M$ by Theorem 2.1.

Now, let λ, P, a, e, and u be as supposed but otherwise arbitrary. Suppose $0 < r \leqq R \leqq a$; let U_R be given by (2.2) with e replaced by $e_R = e$ on $C_R = C(P, R)$ and $e_R = 0$ elsewhere, and define H_R^* by

$$u = U_R + H_R^*$$

Then H_R^* is $\bar{a}$-harmonic on C_R. Clearly

$$l(e_R, C_R) \leqq \begin{cases} L(R/a)^\lambda \cdot (r/R)^\lambda, & 0 \leqq r \leqq R \\ L(R/a)^\lambda, & r \geqq R \end{cases}$$

Hence, from the definition of ϕ, we see that

$$d(u, C_r) \leqq d(U_R, C_r) + d(H_R^*, C_r) \leqq \phi(r/R) \cdot L(R/a)^\lambda + K_7(r/R)^\rho d(H_R^*, C_R)$$

$$\leqq L\phi(r/R) \cdot (R/a)^\lambda + K_7(r/R)^\rho [d(u, C_R) + d(U_R, C_R)]$$

$$\leqq L\phi(r/R) \cdot (R/a)^\lambda + K_7(r/R)^\rho L[\phi(R/a) + M(R/a)^\lambda]$$

using Theorem 2.1 and (1.10). From the arbitrariness of P, a, e, u, and by setting $s = r/a$ and $t = R/a$, we obtain that

$$\phi(s) \leqq t^\lambda \phi(s/t) + K_7(s/t)^\rho \phi(t) + K_7 M s^\rho t^{\lambda - \rho}, \quad 0 < s \leqq t \leqq 1$$

from which the result follows by Lemma 2.1.

LEMMA 2.2. Suppose $\phi(y)$ is continuous and positively homogeneous of degree $1 - \nu$ with $\phi(-y) = -\phi(y)$, $y \neq 0$. Then

$$\int_{C^*(P,a)} \phi(\xi - x)\, d\xi_\alpha' = \int_{C^*(0,1)} \phi(y)\, dy_\alpha', \quad x \in C(P,a)$$

PROOF. This is proved by direct transformation of the surface integrals, the correspondence between ξ and y being given by

$$\xi = x + \lambda y, \quad \lambda > 0, \quad |y| = 1$$

λ being a uniquely determined scalar. The hypothesis $\phi(-y) = -\phi(y)$ is used.

THEOREM 2.3. Suppose e satisfies

(2.12)
$$l(e, E_\nu) \leqq L$$
$$|e(\xi) - e(x)| \leqq L\delta_x^{-\rho-\mu}|\xi - x|^\mu, \qquad |\xi - x| \leqq \delta_x/2$$
$$x \in G, \quad 0 < \mu < 1, \quad \rho = \nu/2$$

and suppose u is given by (2.2). Then

$$|\nabla u(\xi) - \nabla u(x)|$$
$$\leqq K_{11}(\mu, \nu, N, M) \cdot L \cdot \delta_x^{-\rho-\mu}|\xi - x|^\mu, \qquad |\xi - x| \leqq \delta_x/2$$
$$|\nabla u(x)| \leqq K_{12}(\mu, \nu, N, M) \cdot L \cdot \delta_x^{-\rho}, \qquad x \in G$$

The proof of this is entirely classical and is omitted; Lemma 2.2 is useful in the proof.

§3. VARIOUS LEMMAS; CERTAIN OPERATORS

The material in this section is just technical but removes detail from the following section in which the regularity results are presented.

LEMMA 3.1. Suppose that u is of class $\mathfrak{P}_2$ on G and satisfies

$$d[u, C(P, r)] \leqq L(r/\delta_P)^{\rho-1+\mu}$$
$$0 \leqq r \leqq \delta_P, \quad 0 < \mu < 1, \quad \rho = \nu/2$$

for every P in G. Then u is μ-Hölder continuous on G and satisfies

(3.1)
$$|u(\xi) - u(x)| \leqq \mu^{-1}C_1L\delta_x^{1-\rho-\mu}|\xi - x|^\mu, \qquad |\xi - x| \leqq \delta_x/2$$
$$x \in G$$

where C_1 depends only on ν.

Essentially the same lemma has been proved in [12], pp. 13, 14.

DEFINITION 3.1. A function u is said to be <u>in</u> $\mathfrak{P}_{2o}$ on the bounded domain G if and only if it is in $\mathfrak{P}_2$ on G and its average over G is 0. For such u we define $\|u\|^o = d_1(u, G)$. A function u is said to be <u>in</u> H_o on G if and only if u and ∇u are in $\mathfrak{P}_{2o}$ on G; for such u, we define $\|u\|^* = d_2(u, G)$.

DEFINITION 3.2. Given P, a, and λ $(0 \leqq \lambda \leqq \rho)$ we define

$$\|u\|_\lambda^0 = \underset{0<r\leqq a}{\text{l.u.b.}} (r/a)^{-\lambda} d_1[u, C(P, r)]$$

$$\|u\|_\lambda^* = \underset{0<r\leqq a}{\text{l.u.b.}} (r/a)^{-\lambda} d_2[u, C(P, r)]$$

The space $\mathfrak{P}_{2\lambda}$ consists of all (classes of equivalent) u in $\mathfrak{P}_{2o}$ on $C(P, a)$ with $\|u\|_\lambda^o$ finite and the space $H_{o\lambda}$ consists of all u in H_o with $\|u\|_\lambda^*$ finite.

DEFINITION 3.3. If G is a bounded domain, the <u>space</u> C_μ^1 consists of all u in $\mathfrak{P}_{2o}$ and of class C_μ^1 on G with

$$\|u\|_\mu^1 = \max [\|u\|^o, Z_1] \tag{3.2}$$

Z_1 = g.l.b. of all L such that

$$|\nabla u(\xi) - \nabla u(x)| \leqq L\delta_x^{-\rho-\mu} |\xi - x|^\mu$$

$$|\xi - x| \leqq \delta_x/2, \quad x \in G, \quad 0 < \mu < 1$$

Similarly the <u>space</u> C_μ^2 consists of all u in H_o and of class C_μ^2 on G with

$$\|u\|_\mu^2 = \max [\|u\|^*, Z_2]$$

Z_2 = g.l.b. of all L such that

$$|\nabla^2 u(\xi) - \nabla^2 u(x)| \leqq L\delta_x^{-\rho-\mu} |\xi - x|^\mu$$

$$|\xi - x| \leqq \delta_x/2, \quad x \in G, \quad 0 < \mu < 1$$

LEMMA 3.2. If u is in $\mathfrak{P}_{20}$ on $C(P, a)$, then

$$l[u, C(P, a)] \leqq C_2 a \|u\|^{\circ}, \quad C_2 = C_2(\nu)$$

This is just Rellich's theorem for a sphere.

LEMMA 3.3. If u is in $\mathfrak{P}_{2\lambda}$ on $C(P, a)$, then

$$l[u, C(P, r)] \leqq (C_2 + 1) a \|u\|^{\circ}_{\lambda} \cdot (r/a)^{\lambda}$$

$$0 \leqq r \leqq a, \quad 0 \leqq \lambda \leqq \rho$$

PROOF. We prove this for $\lambda \neq \rho - 1$; the result follows for this value by continuity. For $0 \leqq r \leqq a$, define $\bar{u}(r)$ as the average of u over C_r. Then

$$\bar{u}(r) = (\gamma_\nu a^\nu)^{-1} \int_{C(P,a)} U(y, r)\, dy$$

$$U(y, r) = u[x_P + (r/a)(y - x_P)], \quad y \in C_a$$

Then

$$U_r(y, r) = a^{-1}(y^\alpha - x_P^\alpha) u_\alpha [x_P + (r/a)(y - x_P)] = r^{-1}(x^\alpha - x_P^\alpha) u_\alpha(x)$$

Hence, by changing back to x and using Schwarz's inequality, we get

$$\bar{u}'(r) \leqq (\gamma_\nu r^\nu)^{-1/2} d(u, C_r) \leqq (\gamma_\nu r^\nu)^{-1/2} \|u\|^{\circ}_{\lambda} (r/a)^{\lambda}$$

Since $\bar{u}(a) = 0$, we have

$$|\bar{u}(r)| \leqq a \cdot (\gamma_\nu a^\nu)^{-1/2} \|u\|^{\circ}_{\lambda} (\rho - \lambda - 1)^{-1} [(r/a)^{\lambda + 1 - \rho} - 1]$$

Hence, from Lemma 3.2, we see that

$$l(u, C_r) \leqq l[u - u(r), C_r] + l[\bar{u}(r), C_r]$$

$$\leqq C_2 r d(u, C_r) + (\gamma_\nu r^\nu)^{1/2} \bar{u}(r)$$

$$\leqq a \cdot \|u\|^{\circ}_{\lambda} \cdot (r/a)^{\lambda} \left\{ C_2 + (\rho - \lambda - 1)^{-1} [(r/a) - (r/a)^{\rho - \lambda}] \right\}$$

The result follows from an elementary analysis of the second term in the brace.

LEMMA 3.4. If u is in C^1_μ on $C(P, a)$, then

$$|\nabla u(x)| \leqq C_3 \|u\|^1_\mu \delta_x^{-\rho}$$

$$|u(x)| \leqq C_3 C_4 a \|u\|^1_\mu \delta_x^{-\rho}, \qquad x \in C(P, a)$$

$$|u(\xi) - u(x)| \leqq C_3 a \|u\|^1_\mu \delta_x^{-\rho-\mu} |\xi - x|^\mu, \qquad |\xi - x| \leqq \delta_x/2$$

$$d_1[u, C(x, r)] \leqq C_5 \|u\|^1_\mu (r/\delta_x)^\rho$$

$$d_0[u, C(x, r)] \leqq C_6 \|u\|^1_\mu \cdot a \cdot (r/\delta_x)^\rho$$

$$0 \leqq r \leqq \delta_x$$

where C_3, C_4, C_5, C_6 depend only on ν.

PROOF. All these assertions follow from the first and Lemmas 3.2 and 3.3. To prove the first, write $|\nabla u(x)| \leqq |\nabla u(\xi) - \nabla u(x)| + |\nabla u(\xi)|$ and use (3.2) and the Schwarz and Minkowski inequalities.

LEMMA 3.5. If u and ∇u are in $\mathfrak{P}_2$ on the bounded domain G, there is a unique linear function l such that $u - l \in H_0$ on G. If $G = C(P, a)$, we have

$$d_0(l, C_a) \leqq d_0(u, C_a) + C_2^2 a^2 d_2(u, C_a), \quad C_a = C(P, a)$$

$$d_1(l, C_a) \leqq (\nu + 2)^{1/2} a^{-1} d_0(l, C_a)$$

$$d_1(u, C_a) \leqq d_1(l, C_a) + C_2 a d_2(u, C_a)$$

PROOF. The first statement is evident. Write $u = u_0 + l$, $u \in H_0$ on C_a; one application of Lemma 3.2 yields the third inequality, and two applications yield

$$d_0(u_0, C_a) \leqq C_2^2 a^2 d_2(u, C_a)$$

from which the first inequality follows. The second one is easily computed.

DEFINITION 3.4. Suppose f is in L_2 on some bounded set S. By the $\bar{a}$-potential of f, we mean the function V defined by

(3.3) $$V^i(x) = \int_S \Gamma^{ij}(\xi - x) f_j(\xi)\, d\xi$$

LEMMA 3.6. If f and V are as in Definition 3.4, then V and ∇V are in $\mathfrak{P}_2$ on any bounded domain and V satisfies

(3.4) $$\bar{a}^{\alpha\beta}_{ij} V^j_{x^\alpha x^\beta} = f_i$$

almost everywhere and hence satisfies

(3.5) $$\int_{R^*} \bar{a}^{\alpha\beta}_{ij} V^j_{x^\beta}\, dx'_\alpha = \int_R f_i\, dx \qquad (f = 0 \text{ outside } S)$$

on all cells. Moreover, we have

(3.6) $$V^i_{x^\beta}(x) = - \int_S \Gamma^{ij}_\beta(\xi - x) f_j(\xi)\, d\xi$$

Therefore, V is a solution of (2.1) with

(3.7) $$e^\alpha_j(x) = - \delta^{\alpha\beta} f_j(x)$$

PROOF. For smooth f which vanish outside some sphere, (3.4) to (3.7) and the last statement follow. Etc.

DEFINITION 3.5. If f is in L_2 on E_ν and G is a bounded domain, the G-$\bar{a}$-potential of f is $v = V_G - l_G$ where V_G is defined by (3.3) with $S = G$ and l_G is chosen so that v is in H_0 on G. If e is in L_2 on E_ν, the special solution of (2.1) in $\mathfrak{P}_{20}$ on G is that solution of (2.1) with e replaced by e_G, ($e_G = e$ on G and 0 elsewhere) which differs from (2.2) by a constant so chosen that the solution is in $\mathfrak{P}_{20}$ on G.

LEMMA 3.7. Suppose f is in L_2 on $C(P, a)$ and v is its $C(P, a)$-$\bar{a}$-potential. Then v is a solution of (3.4) and (3.5) and v_{x^β} is the special solution of (2.1) in $\mathfrak{P}_{20}$ on $C(P, a)$ with e given by (3.7) on $C(P, a)$ and 0 elsewhere and

$$d_2[v, C(P, a)] \leq \nu^{1/2} M[f, C(P, a)]$$

If f satisfies

(3.8) $$l[f, C(x, r)] \leq L(r/\delta_x)^{\lambda}, \quad 0 \leq r \leq \delta_x,$$

$$0 \leq \lambda < \rho, \quad x \in C_a$$

then

(3.9) $$\|v\|_{\lambda}^{*} \leq \nu^{1/2} K_{10} L$$

If f satisfies

(3.10) $$l[f, C(P, a)] \leq L, \qquad |f(\xi) - f(x)| \leq L\delta_x^{-\rho-\mu} |\xi - x|^{\mu}$$

$$|\xi - x| \leq \delta_x/2, \qquad x \in C_a, \qquad 0 < \mu < 1$$

then v is in C_{μ}^{2} on $C(P, a)$ and

$$\|v\|_{\mu}^{2} \leq \nu^{1/2} K_{13} L, \quad K_{13} = \max [K_{11}, M]$$

If f satisfies (3.8) with $\lambda = \rho - 1 + \mu$, $0 < \mu < 1$, for every x in $C(P, a)$, then v is in C_{μ}^{1} with

$$\|v\|_{\mu}^{1} \leq K_{14} aL, \quad K_{14} = K_{14}(\mu, \nu, N, M)$$

PROOF. The first three statements follow from Lemma 3.6 and Theorems 2.1, 2.2, and 2.3. To prove the last, we write

(3.11) $$v(x) = v_Q(x) + H_Q^{*}(x), \quad x \in C(Q, \delta_Q), \quad Q \in C(P, a)$$

where v_Q is the $C(Q, \delta_Q)$-$\bar{a}$-potential of f. Then H_Q^{*} is $\bar{a}$-harmonic on $C(Q, \delta_Q)$ and

$$d_2[H_Q^{*}, C(Q, \delta_Q)] \leq d_2[v, C(Q, \delta_Q)] + d_2[v_Q, C(Q, \delta_Q)] \leq 2^{1/2} ML$$

by Lemma 3.6 and the first part of Lemma 3.7. From (1.10) we have

(3.12) $$d_2[H_Q^*, C(Q, r)] \leqq 2\nu^{1/2}MK_7L(r/\delta_Q)^{\rho}, \quad 0 \leqq r \leqq \delta_Q$$

Combining (3.12) with (3.9) for $C(Q, \delta_Q)$, we obtain

$$d_2[v, C(Q, r)] \leqq \nu^{1/2}(2K_7M + K_{10}) \cdot L \cdot (r/\delta_Q)^{\rho-1+\mu},$$

$$0 \leqq r \leqq \delta_Q$$

The result follows from Lemmas 3.1 and 3.6.

In the remainder of the section, we assume that $a(x)$, $b(x)$, $c(x)$, and $d(x)$ are bounded and measurable on some sphere $C(P, a)$ and that there exist constants $\bar{a}$ satisfying (1.6) such that

(3.13) $$|a(x) - \bar{a}| \leqq k, \quad |b(x)| \leqq M_2, \quad |c(x)| \leqq M_3$$
$$|d(x)| \leqq M_4, \quad x \in C(P, a)$$

DEFINITION 3.6. For functions u in $\mathfrak{P}_{20}$ on $C(P, a)$, we define the transformation T_1 by $T_1u = U_1 + U_2$ where U_1 is the special solution of (2.1) in $\mathfrak{P}_{20}$ on $C(P, a)$ and U_2 is the $C(P, a)$-$\bar{a}$-potential of f; here

(3.14) $$e = (a - \bar{a}) \cdot \nabla u + b \cdot u, \quad f = c \cdot \nabla u + d \cdot u$$
$$x \in C(P, a)$$

We note that if $U = T_1u$, then $U \in \mathfrak{P}_{20}$ and is a solution of

(3.15) $$\int_{R^*} [\bar{a}_{ij}^{\alpha\beta}U_{x^\beta}^j + (a_{ij}^{\alpha\beta} - \bar{a}_{ij}^{\alpha\beta})u_{x^\beta}^j + b_{ij}^{\alpha}u^j]\, dx'_{\alpha}$$
$$= \int_R (c_{ij}^{\alpha}u_{x^\alpha}^j + d_{ij}u^j)\, dx$$

DEFINITION 3.7. For functions u in H_o on $C(P, a)$, we define T_2u as the $C(P, a)$-$\bar{a}$-potential of $-(a - \bar{a}) \cdot \nabla^2 u + c \cdot \nabla u + d \cdot u$.

THEOREM 3.1. The transformation T_1 is an operator from $\mathfrak{P}_{20}$ to $\mathfrak{P}_{20}$ and from $\mathfrak{P}_{2\lambda}$ to $\mathfrak{P}_{2\lambda}$,

$0 \leqq \lambda < \rho$; T_2 is an operator from H_o to H_o and from $H_{o\lambda}$ to $H_{o\lambda}$, $0 \leqq \lambda < \rho$, and we have

$$\|T_1\|^o \leqq M[k + C_2 a(M_2 + \nu^{1/2}M_3) + \nu^{1/2}C_2^2 a^2 M_4] \tag{3.16}$$

$$\|T_1\|^o_\lambda \leqq K_{10}[k + (C_2 + 1)a(M_2 + \nu^{1/2}M_3) + \nu^{1/2}(C_2 + 1)^2 a^2 M_4] \tag{3.17}$$

$$\|T_2\|^* \leqq \nu^{1/2}M[k + M_3 C_2 a + M_4 C_2^2 a^2] \tag{3.18}$$

$$\|T_2\|^*_\lambda \leqq \nu^{1/2}K_{10}[k + M_3(C_2 + 1)a + M_4(C_2 + 1)^2 a^2] \tag{3.19}$$

PROOF. Let $T_1 u = U_1 + U_2$, $T_2 u = U$ as in the definitions. Setting

$$\begin{gathered} e_1 = (a - \overline{a}) \cdot u, \quad e_2 = b \cdot u, \quad f_1 = -(a - \overline{a}) \cdot \nabla^2 u \\ f_2 = c \cdot \nabla u, \quad f_3 = d \cdot u \end{gathered} \tag{3.20}$$

and using (3.13) and Lemma 3.2, we find that

$$\begin{gathered} l(e_1, C_a) \leqq k\|u\|^o, \quad l(e_2, C_a) \leqq M_2 C_2 a\|u\|^o \\ l(f_2, C_a) \leqq M_3 \|u\|^o, \quad l(f_3, C_a) \leqq M_4 C_2 a\|u\|^o \end{gathered} \tag{3.21}$$

Then (3.16) follows from Theorem 2.1 and Lemmas 3.7 and 3.2. The other results are proved similarly.

THEOREM 3.2. Suppose, in addition to our general hypotheses on a, b, c, d, we assume that a and b satisfy Hölder conditions

$$\begin{gathered} |a(x_1) - a(x_2)| \leqq A_o|x_1 - x_2|^\mu, \quad |b(x_1) - b(x_2)| \leqq B_o|x_1 - x_2|^\mu \\ x_i \in \overline{C}_a, \quad 0 < \mu < 1 \end{gathered} \tag{3.22}$$

Then, if $\overline{a} = a(P)$, T_1 is an operator from C'_μ to C'_μ and

$$\begin{gathered} \|T_1\|^1_\mu \leqq K_{13}[(1 + C_3)A_o a^\mu + C_4 M_2 a + B_o C_3 C_4 a^{1+\mu}] \\ + K_{14}(M_3 C_5 a + M_4 C_6 a^2) \end{gathered} \tag{3.23}$$

If, also, c and d satisfy Hölder conditions like (3.22) with coefficients C_0 and D_0, then T_2 is an operator from C^2_μ to C^2_μ and

$$\|T_2\|^2_\mu \leqq \nu^{1/2}K_{13}[(1 + C_3)A_0 a^\mu + M_3C_4a + M_4C_4a^2 \tag{3.24}$$
$$+ C_oC_3C_4a^{1+\mu} + D_0C_3C_4^2a^{2+\mu}]$$

PROOF. We use the notations of Theorem 3.1. By using Lemma 3.4, letting $e = e_1 + e_2$, $f = f_2 + f_3$, and writing

$$|e_1(\xi) - e_1(x)| \leqq |a(\xi) - \bar{a}| \cdot |\nabla u(\xi) - \nabla u(x)|$$
$$+ |a(\xi) - a(x)| \cdot |\nabla u(x)|$$

etc., we see that e satisfies (2.12) with L replaced by the coefficient of K_{13} in (3.23), and f satisfies (3.8) with L replaced by the coefficient of K_{14} in (3.23). Then (3.23) follows from Theorem 2.3 and Lemma 3.7. The result (3.24) is proved similarly.

§4. REGULARITY PROPERTIES OF THE SOLUTIONS OF (A) AND (C) ON THE INTERIOR

In this section, we retain the general hypotheses of the preceding section on the coefficients $a(x), \ldots, d(x)$ and shall assume in addition that $a(x)$ is continuous on some bounded closed domain $\bar{G}$.

DEFINITION 4.1. By the <u>modulus of continuity of</u> $a(x)$, we mean the function $k(s)$ defined by

$$k(s) = \max |a(x_1) - a(x_2)|, \quad x_1, x_2 \in \bar{G}, \quad |x_1 - x_2| \leqq s$$

LEMMA 4.1. Suppose

$$\omega = (3 - \sqrt{7})/3, \quad \kappa = 3\omega/5 \tag{4.1}$$

There is an integer N_ν such that the closed region $\overline{C(P, 2a\kappa)} - C(P, a\kappa)$ can be covered by N_ν spheres $C[Q, (1 - \omega)\kappa a]$ whose centers Q are at a distance ωa from P.

PROOF. For we see from Analytic Geometry that the circle

$C^*[a\omega, 0; (1-\omega)\kappa a]$ intersects the circles $C^*(0, 0; \kappa a)$ and $C^*(0, 0; 2\kappa a)$ at points $(\kappa a \cos\theta, \pm \kappa a \sin\theta)$ and $(2\kappa a \cos\theta, \pm 2\kappa a \sin\theta)$ respectively where

$$\cos\theta = 9/10$$

THEOREM 4.1. Suppose that e, f, and u are in L_2 on G.

(a) Suppose u is of class $\mathfrak{P}_2$ on domains interior to G and satisfies (A) on G. Then there are numbers $d_0 > 0$, K_{15}, K_{16}, and K_{17} which depend only on ν, N, M, M_2, M_3, M_4, and the modulus of continuity of $a(x)$ on $\bar{G}$ such that

$$d[u, C(P, \kappa a)] \leqq a^{-1}\left\{K_{15} l[u, C(P, a)] + K_{16} l[e, C(P, a)] + K_{17} l[f, C(P, a)]\right\} \tag{4.2}$$

for all P in G, $a \leqq d_0$, κ being defined by (4.1).

(b) Suppose u and ∇u are of class $\mathfrak{P}_2$ on domains interior to G and u satisfies (C) on G. Then there are numbers $d_1 > 0$, K_{18}, and K_{19}, depending only on the quantities above (without M_2) such that

$$d_2[u, C(P, \kappa a)] \leqq a^{-2}\left\{K_{18} l[u, C(P, a)] + K_{19} l[f, C(P, a)]\right\} \tag{4.3}$$

$$\text{all } P \in G, \quad a \leqq d_1$$

PROOF. The proof of both parts follows the method of Schauder [13]. In proving (a), we may as well assume that u is of class $\mathfrak{P}_2$ on G since that is true for any interior domain.

Define

$$k_\nu = [1 + N_\nu (1-\omega)^{-2}]^{1/2} \tag{4.4}$$

N_ν and ω being the numbers of Lemma 4.1. It is clear that there is a number $d_0 > 0$ depending only on the quantities indicated such that

$$k_\nu(1 + C_2K_8)\epsilon(2\kappa d_0) \leqq \frac{1}{2} \tag{4.5}$$

where $\epsilon(a)$ is the right side of (3.16) with k replaced by $k(a)$. For all P in $\overline{G}$ and all a with $0 \leqq a \leqq \delta_P$ and d_0, define

$$w(P, a) = a \cdot d[u, C(P, \kappa a)], \quad P \in G, \quad w(P, 0) = 0, \quad P \in G^* \tag{4.6}$$

Then $w(P, a)$ is continuous in its arguments on the indicated compact set and so takes on its maximum w at some point (P_0, a_0) where, obviously $P_0 \in G$ and $0 < a_0 \leqq \delta_{P_0}$ and d_0.

Clearly we have

$$d[u, C(P_0, \kappa a_0)] = a_0^{-1} w \tag{4.7}$$

Also, each P at a distance ωa_0 from P_0 is at a distance $\geqq (1 - \omega)a_0$ from G^* and, of course, $(1 - \omega)a_0 \leqq d_0$ so that

$$d\left\{u, C[P, (1 - \omega)\kappa a_0]\right\} \leqq (1 - \omega)^{-1} a_0^{-1} w \tag{4.8}$$

From Lemma 4.1, we can cover $\overline{C(P_0, 2\kappa a_0)} - C(P_0, \kappa a_0)$ by N_ν of the spheres (4.8). By squaring, adding, and taking the square root, we obtain, see (4.4),

$$d[u, C(P_0, 2\kappa a_0)] \leqq k_\nu a_0^{-1} w \tag{4.9}$$

Now, let $R = 2\kappa a_0$, $r = \kappa a_0$, $C_s = C(P_0, s)$, and let $\overline{u}$ be the average of u over C_R. Let $u_0 = u - \overline{u} \in \mathfrak{P}_{20}$ on C_R and obviously

$$l(\overline{u}, C_R) \leqq l(u, C_R) \tag{4.10}$$

We note that u_0 satisfies

$$\int_{R^*} [\overline{a}_{ij}^{\alpha\beta} u_{0x\beta}^j + (a_{ij}^{\alpha\beta} - \overline{a}_{ij}^{\alpha\beta}) u_{0x\beta}^j + b_{ij}^{\alpha} u_0^j + (b_{ij}^{\alpha} \overline{u}^j + e_i^{\alpha})]\, dx'_\alpha = \int_R (c_{ij}^{\alpha} u_{0x\alpha}^j + d_{ij} u_0^j + d_{ij} \overline{u}^j + f_i)\, dx \tag{4.11}$$

$$\overline{a} = a(P_0)$$

on C_R. We define $U = T_1 u_0$ (T_1 on C_R), V_1 as the special solution of (2.1) in $\mathfrak{P}_{20}$ on C_R with e replaced by $b \cdot \overline{u} + e$, and V_2 as the C_R-$\overline{a}$-potential of $d \cdot \overline{u} + f$. Then, using (4.9), (4.10), Theorems 2.1 and

3.1, and Lemma 3.7, we compute

$$d(U, C_R) \leqq k_\nu \epsilon(R) \cdot a_0^{-1} w$$

$$d(V, C_R) \leqq Z_1 l(u, C_R) + Z_2 l(e, C_R) + Z_3 l(f, C_R) = Z_4, \tag{4.12}$$

$$V = V_1 + V_2$$

where Z_1, Z_2, Z_3 depend only on the quantities indicated. From (4.11), we have

$$u = U + V + H^* \tag{4.13}$$

where H^* is $\bar{a}$-harmonic.

Now, from (4.12) and Lemma 3.2, we have

$$l(U, C_R) \leqq C_2 R k_\nu \epsilon(R) a_0^{-1} w, \quad l(V, C_R) \leqq C_2 R Z_4$$

so that

$$l(H^*, C_R) \leqq l(u, C_R) + C_2 R[k_\nu \epsilon(R) a_0^{-1} w + Z_4]$$

From (1.10), we conclude that

$$\begin{aligned} d(H^*, C_R) &\leqq K_8 R^{-1} l(H^*, C_R) \\ &\leqq K_8 C_2 k_\nu \epsilon(R) a_0^{-1} w + K_8 C_2 Z_4 + K_8 (2\kappa)^{-1} a_0^{-1} l(u, C_R) \end{aligned} \tag{4.14}$$

From (4.7), (4.12), (4.13), and (4.14), we conclude that

$$\begin{aligned} a_0^{-1} w \leqq\ & k_\nu (1 + K_8 C_2) \epsilon(R) a_0^{-1} w + (1 + K_8 C_2) Z_4 \\ & + K_8 (2\kappa)^{-1} a_0^{-1} l(u, C_R) \end{aligned} \tag{4.15}$$

Since the coefficient of $a_0^{-1} w$ on the right in (4.15) is $\leqq 1/2$ by (4.5), we obtain

$$w \leqq 2 d_0 (1 + K_8 C_2) Z_4 + 2 K_8 (2\kappa)^{-1} l(u, C_R)$$

which proves (a).

The proof of (b) is entirely similar. We assume u and ∇u in

$\mathfrak{B}_2$ on G and define $w(P, a)$ as in (4.6) except

$$w(P, a) = a^2 d_2[u, C(P, \kappa a)]$$

Lemma 3.5 is useful.

THEOREM 4.2. Suppose e and f satisfy (2.11), $C(P, a) \subset G$.

(a) Suppose u is of class $\mathfrak{B}_2$, and satisfies (A) on $C(P, a)$. Then there are numbers $d_2 > 0$, K_{20}, and K_{21} depending only on λ and the quantities in Theorem 4.1 such that

$$d[u, C(P, r)]$$

(4.16)
$$\leq \left\{3K_7 d[u, C(P, a)] + K_{20} l[u, C(P, a)] + K_{21} L\right\}(r/a)^{\lambda}$$

$$0 \leq r \leq a \leq d_2$$

If $\lambda = \rho - 1 + \mu$, $0 < \mu < 1$, and e and f satisfy (2.11) with $C(P, a)$ replaced by any $C(x, \delta_x) \subset C(P, a)$, then u satisfies a condition (3.1) on $C(P, a)$ with L replaced by the coefficient in (4.16).

(b) Suppose u and ∇u are of class $\mathfrak{B}_2$ on $C(P, a)$ and u satisfies (C) there. Then there are numbers $d_3 > 0$, K_{22}, K_{23}, and K_{24} depending only on the quantities above (except M_2) such that

$$d_2[u, C(P, r)]$$

(4.17)
$$\leq \left\{ K_{22} d_2[u, C(P, a)] + K_{23} a^{-1} d_0[u, C(P, a)] + K_{24} L \right\}(r/a)^{\lambda}$$

$$0 \leq r \leq a \leq d_3$$

If the additional hypothesis in (a) holds then u is of class C'_μ on $C(P, a)$ and ∇u satisfies (3.1) on $C(P, a)$ with L replaced by the coefficient in (4.17).

PROOF. The proof of (b) is similar to that of (a) and will be omitted. To prove (a) we choose $d_2 > 0$ but so small that the right sides of (3.16) and (3.17) are $\leqq 1/2$ whenever $a \leqq d_2$, $k = k(a)$. Suppose $a \leqq d_2$, $C(P, a) \subset G$, and e and f satisfy (2.11) at P. We define u_o, $\overline{u}$, U, V, and H^* as in the proof of Theorem 4.1 with $C(P_o, R)$ replaced by $C(P, a)$. Then (4.11) and (4.13) hold and H^* is $\overline{a}$-harmonic on $C(P, a)$. If we let $e^* = b \cdot \overline{u} + e$, $f^* = d \cdot \overline{u} + f$, we see that e^* and f^* satisfy (2.11) at P with L replaced by

$$(4.18) \quad Z_1 = L + M_2 l(u, C_a), \quad Z_2 = L + M_4 l(u, C_a), \quad C_s = C(P, s)$$

respectively. Since $\|T_1\|^o \leqq 1/2$, $\|T_1\|^o_\lambda \leqq 1/2$, we obtain

$$(4.19) \quad \begin{aligned} &\|U\|^o_\lambda \leqq \|u\|^o_\lambda/2, \quad \|V\|^o_\lambda \leqq K_{10}[Z_1 + \nu^{1/2}(C_2 + 1)aZ_2] \\ &d(U, C_a) \leqq \|u\|^o/2, \quad d(V, C_a) \leqq M(Z_1 + \nu^{1/2}C_2 aZ_2] \end{aligned}$$

using (4.18), Theorems 2.1 and 2.2, and some lemmas. From (4.13), (4.18), (4.19), and (1.10), we obtain

$$(4.20) \quad \begin{aligned} &d(H^*, C_a) \leqq 3d(u, C_a)/2 + d(V, C_a) \\ &d(H^*, C_r) \leqq K_7 d(H^*, C_a) \cdot (r/a)^\lambda \\ &0 \leqq r \leqq a \end{aligned}$$

Since $\|U\|^o_\lambda \leqq \|u\|^o_\lambda/2$, the result follows from (4.11), (4.19), and (4.20). The last statement of Theorem 4.2 follows from this and Lemma 3.1.

THEOREM 4.3. *Suppose e and f are in L_2 on $C(P, a)$ and satisfy*

$$(4.21) \quad \begin{aligned} &l(e, C_a) \leqq L \\ &|e(\xi) - e(x)| \leqq L\delta_x^{-\rho-\mu}|\xi - x|^\mu, \qquad |\xi - x| \leqq \delta_x/2 \\ &l[f, C(x, r)] \leqq L(r/\delta_x)^{\rho-1+\mu}, \qquad 0 \leqq r \leqq \delta_x, \qquad 0 < \mu < 1 \\ &x \in C_a = C(P, a) \end{aligned}$$

Suppose u is of class $\mathfrak{P}_2$ and satisfies (A) on C_a. Suppose also that a and b satisfy (3.22) on $\overline{C}_a$. Then u is of class C'_μ on C_a and there are numbers $d_4 > 0$, K_{25}, K_{26}, and K_{27}

depending only on the quantities in Theorem 4.2 and A_o and B_o such that for $a \leqq d_4$

$$|\nabla u(\xi) - \nabla u(x)| \leqq [K_{25} d(u, C_a) + K_{26} l(u, C_a) + K_{27} L] \cdot \delta_x^{-\rho-\mu} |\xi - x|^{\mu}$$

$$|\xi - x| \leqq \delta_x/2, \quad \text{etc.}$$

PROOF. We may choose d_4 so small that the right sides of (3.16) and (3.23) are $\leqq 1/2$ for $a \leqq d_4$, $k = A_o a^{\mu}$. Using the notations and methods of proof in Theorem 4.2 and using (1.10), we obtain

(4.22)
$$d(H^*, C_a) \leqq \tfrac{3}{2} d(u, C_a) + M[Z_1 + \nu^{1/2} Z_2]$$
$$\|H^*\|_{\mu}^{1} \leqq \tfrac{1}{3} K_{25} d(H^*, C_a), \quad K_{25} = \max[3, 3 \cdot 2^{1+\rho} K_5]$$

Using (4.10), etc., we see that e^* and f^* satisfy (4.21) with L replaced by

(4.23)
$$Z_3 = L + Z_5 l(u, C_a), \quad Z_4 = L + M_4 l(u, C_a)$$
$$Z_5 = \max[\gamma_{\nu}^{-1/2} B_o a^{\mu}, M_2]$$

Hence from Theorem 2.3, Lemma 3.7, etc., we obtain

$$\|V\|_{\mu}^{1} \leqq K_{13} Z_3 + K_{14} a Z_4$$

Since $\|U\|_{\mu}^{1} \leq \|u\|_{\mu}^{1/2}$, the result follows.

We now summarize the regularity results for equations (A) so far obtained in this section in the following theorem:

THEOREM 4.4. Suppose the coefficients $a, b, \ldots, f$ satisfy the fundamental hypotheses of this section. Suppose u is in L_2 on G, of class $\mathfrak{P}_2$ on interior domains, and satisfies (A) on G. Then

(4.24)
$$d(u, G_1) \leqq L_1, \quad \overline{G}_1 \subset G$$

where L_1 depends only on ν, N, M, M_2, M_3, M_4, G_1, $l(u, G)$, $l(e, G)$, $l(f, G)$, the modulus of continuity of $a(x)$, and the distance of G_1 from G^*.

If e and f satisfy (2.11) with $\lambda = \rho - 1 + \mu$, $0 < \mu < 1$, for all $C(P, a) \subset G$, then u is continuous on G and satisfies

(4.25) $$|u(x_1)| \leq L_2, \qquad |u(x_1) - u(x_2)| \leq L_3|x_1 - x_2|^{\mu}$$
$$x_k \in \overline{G}_1 \subset G$$

where L_2 and L_3 depend on the quantities above and μ and L.

If, in addition, a and b satisfy (3.22) and e and f satisfy (4.21) for every $C(P, a) \subset G$, then u is of class C'_{μ} on G and satisfies

(4.26) $$|\nabla u(x_1)| \leq L_4, \quad |\nabla u(x_1) - \nabla u(x_2)| \leq L_5|x_1 - x_2|^{\mu}$$
$$x_k \in \overline{G}_1 \subset G$$

where L_4 and L_5 depend on the quantities above and A_0 and B_0.

PROOF. (4.24) follows since $\overline{G}_1$ can be covered by a finite number of spheres $C(P, \kappa a_P)$, $P \in G$, where a_P is the smaller of d_0 and δ_P and κ is the number of Lemma 4.1. The other results may be obtained by first interposing a domain G_2 between G_1 and G on which (4.24) holds and then covering $\overline{G}_1$ by a finite number of $C(P, a) \subset G_2$ with a small enough. The bound for $|\nabla u(x_1)|$ may be obtained using Lemma 3.4.

We now employ the device of Lichtenstein [9] to prove the following further theorem:

THEOREM 4.5. Suppose, in addition to all the hypotheses of Theorem 4.4, that c, d, and f satisfy (3.22) and that a, b, and e are of class C'_{μ} on $\overline{G}$ satisfying

(4.27) $$|\nabla a(x_1) - \nabla a(x_2)| \leq A_1|x_1 - x_2|^{\mu}, \text{ etc.}$$
$$x_k \in \overline{G}, \qquad 0 < \mu < 1$$

Then u is of class C''_{μ} on G and satisfies

$$(4.28)\quad |\nabla^2 u(x_1)| \leq L_6, \qquad |\nabla^2 u(x_1) - \nabla^2 u(x_2)| \leq L_7 |x_1 - x_2|^\mu$$

$$x_k \in \overline{G}_1 \subset G$$

where L_6 and L_7 depend only on the previous quantities and on A_1, B_1, E_1, C_0, D_0, and F_0.

PROOF. Suppose $\overline{G}_1 \subset G$. Choose domains G_2 and G_3 so $\overline{G}_1 \subset G_2 \subset \overline{G}_2 \subset G_3 \subset \overline{G}_3 \subset G$. Then, by Theorem 4.4, u is of class C'_μ on $\overline{G}_3$ and satisfies (4.24) ... (4.26) on $\overline{G}_3$. Let v be the (ordinary) G_3-potential of $c \cdot \nabla u + d \cdot u + f$. If G_3 is smooth, then v is of class C^2_μ on $\overline{G}_3$ and satisfies conditions like (4.24) ... (4.28) there. Hence u satisfies

$$(4.29)\qquad \int_{R^*} (a^{\alpha\beta}_{ij} u^j_{x^\beta} + b^\alpha_{ij} u^j + e^\alpha_i - v^i_{x^\alpha})\, dx'_\alpha = 0$$

on G_3.

We now apply the device of Lichtenstein to (4.29) as follows: Let $h_0 > 0$ be the distance of G_2 from G_3^* and let ζ be any unit vector. For each h with $0 < |h| < h_0$, define

$$p(x) = h^{-1}[u(x + h\zeta) - u(x)], \qquad x \in \overline{G}_2$$

$$\begin{aligned} e_1(x) = b(x + h\zeta) \cdot p(x) + h^{-1} \{ & [a(x + h\zeta) - a(x)] \cdot \nabla u(x) \\ & + [b(x + h\zeta) - b(x)] \cdot u(x) \\ & + [e(x + h\zeta) - e(x)] \\ & - [\nabla v(x + h\zeta) - \nabla v(x)]\} \end{aligned}$$

By subtracting (4.29) from (4.29) with x replaced by $x + h\zeta$, we see that $p(x)$ satisfies

$$\int_{R^*} [a^{\alpha\beta}_{ij}(x + h\zeta) p^j_{x^\beta}(x) + e^\alpha_{1i}(x)]\, dx'_\alpha = 0$$

on G_2. Moreover, from (4.24) ... (4.27), we see that p is of class C'_μ on $\overline{G}_2$ with

$$|p(x_1)| \leq L_4, \qquad |p(x_1) - p(x_2)| \leq L_5 |x_1 - x_2|^\mu,$$

$$x_k \in \overline{G}_2$$

$$|e(x_1)| \leq Z_1, \qquad |e(x_1) - e(x_2)| \leq Z_2 |x_1 - x_2|^\mu$$

where all the bounds are independent of h, Z_1 and Z_2 and depend only on h_o and all the other quantities. From Theorem 4.4, it follows that there are numbers Z_3 and Z_4 depending only on these quantities such that

$$|\nabla p(x_1)| \leqq Z_3, \quad |\nabla p(x_1) - \nabla p(x_2)| \leqq Z_4 |x_1 - x_2|^{\mu}, \quad x_k \in \overline{G}_1$$

independently of h. The results follow since we may then let $h \longrightarrow 0$.

If the hypotheses of Theorem 4.5 hold so that u is of class C''_{μ} on G, then u satisfies (D) which is a special case of (C). Concerning the system (C) we can first prove Theorem 4.6 below by the methods of proof of Theorems 4.2 and 4.3, and then we can prove Theorem 4.7 using Theorem 4.6 and the device of Lichtenstein repeatedly.

THEOREM 4.6. Suppose, in addition to our general hypotheses, we assume that a, c, and d satisfy (3.22) on $C(P, a)$ and f satisfies (2.12) on $C(P, a)$. Suppose that u and ∇u are in $\mathfrak{P}_2$ on $C(P, a)$ and u satisfies (C) there. Then u is of class C''_{μ} on $C(P, a)$ and there are numbers $d_5 > 0$, K_{28}, K_{29}, K_{30}, depending only on μ, ν, N, M, M_3, M_4, A_o, C_o, and D_o such that

$$|\nabla^2 u(\xi) - \nabla^2 u(x)|$$

$$\leqq [K_{28} d_2(u, C_a) + a^{-1} K_{29} d_o(u, C_a) + K_{30} L] \delta_x^{-\rho-\mu} \, |\xi - x|^{\mu}$$

$$|\xi - x| \leqq \delta_x/2, \quad x \in C(P, a), \quad \text{if} \quad a \leqq d_5$$

THEOREM 4.7. Suppose f is in L_2 on the bounded domain G and $a(x)$, $c(x)$, and $d(x)$ satisfy the fundamental hypotheses of this section. Suppose that u is in L_2 on G, that u and ∇u are in $\mathfrak{P}_2$ on interior domains, and that u satisfies (C) on G. Then

$$d_2(u, G_1) \leqq L_1, \quad \overline{G}_1 \subset G$$

where L_1 depends only on ν, N, M, M_3, M_4, G_1, $l(u, G)$, $l(f, G)$, the modulus of continuity of $a(x)$, and the distance of G from G^*.

If f satisfies (2.11) with $\lambda = \rho - 1 + \mu$, $0 < \mu < 1$, for all $C(P, a) \subset G$, then u is of class C'_μ on G and satisfies

$$|u(x_1)| \leqq L_2, \quad |\nabla u(x_1)| \leqq L_3$$

$$|\nabla u(x_1) - \nabla u(x_2)| \leqq L_4 |x_1 - x_2|^\mu, \quad x_k \in \overline{G}_1 \subset G$$

where L_2, L_3, and L_4 depend only on the quantities above and on μ and L.

If a, c, d, and f are of class $C_\mu^{(n)}$ on $\overline{G}$, then u is of class $C_\mu^{(n+2)}$ on G; if $\overline{G}_1 \subset G$, the bounds for $\nabla^k u$, $(0 \leqq k \leqq n + 2)$ and the Hölder coefficient for $\nabla^{n+2} u$ on $\overline{G}_1$ depend only on the quantities above, on the bounds for the coefficients and their derivatives up to order n, and on the Hölder coefficients for the n-th derivatives of a, c, f and d on G.

§5. FURTHER PRELIMINARY LEMMAS

We begin by defining certain sets and quantities as follows:

DEFINITION 5.1. Let E_ν^+ denote the part of E_ν where $x^\nu > 0$ and σ denote the set $x^\nu = 0$. We denote $C(P, a) \cap E_\nu^+$ by $G(P, a)$, abbreviating $G(P, a)$ to G_a and $C(P, a)$ to C_a if P is the origin. We denote $C_a \cap \sigma$ by σ_a. If $P \in C_a$ or G_a, d_P denotes the distance of P from C_a^*, not G_a^*. If u is of class $\mathfrak{P}_2$ on G_a, we say that it vanishes on σ_a if and only if u is of class $\mathfrak{P}_2$ on C_a when extended to C_a by reflection, i.e., by

$$u(-x^\nu, x'_\nu) = -u(x^\nu, x'_\nu) \tag{5.1}$$

LEMMA 5.1. Suppose u is of class $\mathfrak{P}_2$ on the bounded domain G and vanishes on G^*. Then

$$I(u, G) \leqq a d(u, G), \quad \gamma_\nu a^\nu = m(G)$$

This is an immediate consequence of [11], Theorem 9.3.

LEMMA 5.2. If u is of class $\mathfrak{P}_2$ on G_a and vanishes on σ_a, then

$$l(u, G_a) \leqq ad(u, G)$$

The proof is immediate.

LEMMA 5.3. Suppose u is in $\mathfrak{P}_2$ on the bounded domain G and H is the harmonic function coinciding with u on G^*. Then

$$l(H, G) \leqq l(u, G) + ad(u, G)$$

$$l(u, G) \leqq l(H, G) + ad(u, G)$$

$$\gamma_\nu a^\nu = m(G)$$

PROOF. Set $u = U + H$, $U = 0$ on G^*, $d(U, G) \leqq d(u, G)$. Use Lemma 5.1.

LEMMA 5.4. Suppose u is of class $\mathfrak{P}_2$ on the bounded domain G. There is a unique linear function $l = l_\alpha x^\alpha$ which minimizes $d(u - l, G)$ among all such l. If H is the harmonic function coinciding with u on G^*, the corresponding linear function for H is also l. If $G = C(P, a)$, l differs by a constant from the linear function tangent to H at P.

PROOF. The first statement is evident and the second follows since, for any such l, we have

$$D_2(u - l, G) = D_2(H - l, G) + D_2(u - H, G)$$

LEMMA 5.5. Suppose u is of class $\mathfrak{P}_2$ on $G(P, a)$ and

$$d[u - l_{P,r}, G(P, r)] \leqq L(r/a)^{\rho+\mu}, \tag{5.2}$$

$$0 \leqq r \leqq a, \quad 0 < \mu < 1, \quad \rho = \nu/2$$

where $l_{P,r}$ is the function of Lemma 5.4 for u on $G(P, r)$. Then there is a unique homogeneous linear function l_P such that

$$d[u - l_P, G(P, r)] \leqq C_7 \cdot L(r/a)^{\rho+\mu}, \quad C_7 = C_7(\mu, \nu) \tag{5.3}$$

The generalized derivatives of u exist at P and $\nabla u(P) = \nabla l_P$.

PROOF. Writing (5.2) for $0 < s \leq r \leq a$, we obtain

$$d[u - l_{P,s}, G(P, s)] \leq L(s/a)^{\rho+\mu}$$

$$d[u - l_{P,r}, G(P, s)] \leq r/s)^{\rho+\mu} \cdot L(s/a)^{\rho+\mu}$$

since $G(P, s) \subset G(P, r)$. Squaring subtracting and then taking square roots, we obtain (after dividing by $\{m[G(P, s)]\}^{1/2}$)

$$|\nabla l_{P,r} - \nabla l_{P,s}| \leq (2/\gamma_\nu)^{1/2}[1 + (r/s)^{\rho+\mu}] \cdot La^{-\rho-\mu}s^{\mu}$$

(5.4)

$$m[G(P, s)] \geq \gamma_\nu s^{\nu/2}$$

Now, choose $0 < h < 1$, $r_n = ah^n$, $l_n = l_{P,r_n}$. From (5.4), we obtain

$$|\nabla l_{n+1} - \nabla l_n| \leq Z_1 La^{-\rho-\mu}(a^\mu h^{(n+1)\mu}) = Z_1 La^{-\rho-\mu} r^{\mu}_{n+1}, \qquad n,p = 1, 2, \ldots,$$

(5.5)

$$|\nabla l_{n+p} - \nabla l_n| \leq (1 - h^\mu)^{-1} Z_1 La^{-\rho-\mu} r^{\mu}_{n+1}, \qquad Z_1 = (2/\gamma_\nu)^{1/2}(1 + h^{-\rho-\mu})$$

From (5.5) we conclude that $\nabla l_n \longrightarrow$ some ∇l_P and that we may let $p \longrightarrow \infty$ in (5.5). Using (5.4) for $r_{n+1} < r \leq r_n$, we find that

$$|\nabla l_{P,r} - \nabla l_P|$$

(5.6)

$$\leq (2/\gamma_\nu)^{1/2}(1 + h^{-\rho-\mu}) \cdot [1 + (1 - h^\mu)^{-1}] La^{-\rho-\mu} r^\mu, \quad 0 \leq r \leq a$$

Since (5.6) holds for all h, $0 < h < 1$, we may take the minimizing h. The result follows easily by combining the resulting (5.6) with (5.2).

LEMMA 5.6. Suppose that u is of class $\mathfrak{P}_2$ on G_a and satisfies (5.2) on $G(P, d_P)$ for every $P \in G_a \cup \sigma_a$. Then u is of class C'_μ on $G_a \cup \sigma_a$ and

$$(5.7)\qquad \begin{gathered} |\nabla u(\xi) - \nabla u(x)| \leqq C_8 \, L \, d_x^{-\rho-\mu} |\xi - x|^{\mu} \\ |\xi - x| \leqq d_x/2, \qquad \xi, x \in G_a \cup \sigma_a \\ C_8 = 2^{\rho+\mu+3/2} \gamma_\nu^{-1/2} \quad C_7 = C_8(\mu, \nu) \end{gathered}$$

PROOF. Suppose $x, \xi \in G_a \cup \sigma_a$, $|\xi - x| = r \leqq d_x/2$. By Lemma 5.5, $\nabla u(x)$ and $\nabla u(\xi)$ both exist. Let l_x and l_ξ be the corresponding homogeneous linear functions. From Lemma 5.5, we obtain

$$d[u - l_\xi, G(\xi, r)] \leqq C_7 L (r/d_\xi)^{\rho+\mu} \leqq 2^{\rho+\mu} C_7 L (r/d_x)^{\rho+\mu}$$

$$d[u - l_x, G(\xi, r)] \leqq d[u - l_x, G(x, 2r)] \leqq 2^{\rho+\mu} C_7 L (r/d_x)^{\rho+\mu}$$

$$d[l_x - l_\xi, G(\xi, r)] \leqq 2^{\rho+\mu+1} C_7 (r/d_x)^{\rho+\mu}$$

The result follows by dividing by

$$\{m[G(\xi, r)]\}^{1/2} \geqq (\gamma_\nu r^\nu/2)^{1/2}$$

LEMMA 5.7. Suppose u is of class $C^{(n+1)}$ with $\nabla^k u$ in L_2 on $G(P, a)$, $k \leqq n + 1$, $n \geqq [\rho]$. Then

$$(5.8)\qquad |u(x)| \leqq \sqrt{\frac{2}{\gamma_\nu}} \frac{1}{a^\rho} \left\{ \sum_{j=0}^{n} \frac{(2a)^j}{j!} d_j[u, G(P, a)] + \frac{2(2a)^{n+1}}{n!} d_{n+1}[u, G(P, a)] \right\}$$

PROOF. By expanding in a Taylor series with remainder about each y in $G(P, a)$ and integrating with respect to y, we obtain

$$(5.9)\qquad \begin{aligned} m[G(P, a)] \cdot u(x) = {} & \sum_{j=0}^{n} \frac{(-1)^j}{j!} \int_{G(P,a)} \nabla^j u(y) \cdot (y - x)^j \, dy \\ & + \int_0^1 \frac{(-1)^{n+1}}{n!} t^n \, dt \int_{G(P,a)} \nabla^{n+1} u[x + t(y - x)] \cdot (y - x)^{n+1} \, dy \end{aligned}$$

The inequality follows by applying the Schwarz inequality to each term, setting $z = x + t(y - x)$ to handle the integral of $|\nabla^{n+1} u[x + t(y - x)]|^2$.

LEMMA 5.8. Suppose u is of class C' on $\overline{C(P, R)}$ and H is the harmonic function coinciding with u on $C^*(P, R)$. Then

$$D_2[H, C(P, R)]$$
$$\leq \left[\int_{C^*(P,R)} (|\nabla u|^2 - |u_r^2|)\, dS \right]^{1/2} \cdot \left[\int_{C^*(P,R)} |u|^2\, dS \right]^{1/2}$$

PROOF. If either factor on the right vanishes, the lemma holds, since H is a constant in either case. So suppose neither vanishes. Define

$$A = \left[R^{-1} \int_{C^*(P,R)} |u|^2\, dS \right]^{1/2}, \quad B = \left[R \int_{C^*(P,R)} (|\nabla u|^2 - |u_r^2|)\, dS \right]^{1/2}$$

(5.10)

$$w(r, p) = u(R, p) \cdot (r/R)^{\alpha}, \quad \alpha = B/A, \quad 0 \leq r \leq R$$

where p denotes a point on the unit sphere and (r, p) are polar coordinates with pole at P. Then

$$D_2[H, C(P, R)] \leq D_2[w, C(P, R)]$$
$$= \frac{\alpha^2 A^2 + B^2}{\nu + 2\alpha - 2} = \frac{(\alpha A - B)^2 + 2\alpha AB}{\nu - 2 + 2\alpha} \leq AB$$

since $\nu \geq 2$, $\alpha > 0$, and $A\alpha = B$ by (5.10).

Essentially the following theorem was proved by Shiffman [15].

THEOREM 5.1. Suppose u is in L_2 on $C(P, a)$ and in $\mathfrak{P}_2$ on $C(P, r)$ for any $r < a$. Suppose also that

(5.11) $$D_2[u, C(P, r)] \leq K \cdot D_2[H_r, C(P, r)] + X(r), \quad 0 < r < a$$

where X is summable on $[0, a]$ and H_r denotes the harmonic function which coincides with u on $C^*(P, r)$. Then

$$D_2[u, C(P, r)]$$

(5.12) $$\leqq 4K^2(a - r)^{-2}L_2[u, C(P, a)] + 2(a - r)^{-1} \int_0^a \chi(r)\, dr$$

$$0 < r < a$$

PROOF. For $0 \leqq r < a$, define

$$\phi(r) = D_2(u, C_r), \qquad \psi(r) = L_2(u, C_r)$$

$$\Phi(r) = \int_0^r \phi(s)\, ds, \quad X(r) = \int_0^r \chi(s)\, ds$$

$$C_r = C(P, r)$$

If u were of class C' in $C(P, a)$, we could conclude from Lemma 5.8 that

$$D_2(H_r, C_r) \leqq [\Phi'(r)]^{1/2} \cdot [\psi'(r)]^{1/2}, \quad 0 \leqq r < a$$

and then by integrating and using Schwarz, we would get

(5.13) $$\int_0^r D_2(H_s, C_s)\, ds \leqq [\phi(r)]^{1/2}[\psi(r)]^{1/2} = [\Phi'(r)]^{1/2} \cdot [\psi(r)]^{1/2}$$

$$0 \leqq r < a$$

But now an easy approximation using the facts that the h-average function u_h is of class C' on $C(P, a - \nu^{1/2}h)$ and $u_h \rightrightarrows u$ in $\mathfrak{P}_2$ on any C_r, etc., yields (5.13) in general. Integrating (5.11) and using (5.13), we obtain

(5.14) $$\begin{aligned}\Phi(r) &\leqq K[\Phi'(r)]^{1/2} \cdot [\psi(r)]^{1/2} + X(r) \\ &\leqq K[\psi(a)]^{1/2} \cdot [\Phi'(r)]^{1/2} + X(a), \qquad 0 < r < a\end{aligned}$$

Φ being of class C' and convex. One sees easily that (5.14) implies

(5.15) $$\Phi(r) \leqq X(a) + K^2(a - r)^{-1}\psi(a), \quad 0 \leqq r < a$$

from which (5.12) follows easily.

§6. STRONGLY ELLIPTIC EQUATIONS (1.1) AND (2.1) ON BOUNDED DOMAINS

In this and the following sections we assume that $a(x)$ is continuous on $\overline{G}$ (compact) and satisfies

$$(6.1) \qquad |\lambda|^2 \cdot |\xi|^2 \leqq a_{ij}^{\alpha\beta}(x)\lambda_\alpha\lambda_\beta\xi^i\xi^j, \qquad |a(x)| \leqq M_1$$

b, c and d will satisfy their previous conditions. It is clear that (6.1) represents no loss in generality since the strong elliptic condition (E) and the continuity of the functions in (6.1) on the compact set $x \in \overline{G}$, $|\lambda| = |\xi| = 1$ guarantees a positive minimum for the function on the right there. For our systems, the previous M evidently depends only on ν, N, and M_1 so that all of our previous results hold. In this section, we shall assume that $a(x)$ is a constant tensor $\overline{a}$.

LEMMA 6.1. Suppose u is in $\mathfrak{P}_2$ and vanishes on G^*. Then

$$\int_G \overline{a}_{ij}^{\alpha\beta} u_{x^\alpha}^i u_{x^\beta}^j \, dx \geqq D_2(u, G)$$

PROOF. The integral is unaltered if $\overline{a}$ is replaced by

$${}'\overline{a}_{ij}^{\alpha\beta} = \frac{1}{2}(\overline{a}_{ij}^{\alpha\beta} + \overline{a}_{ji}^{\beta\alpha})$$

and ${}'\overline{a}$ still satisfies (6.1). The result for ${}'\overline{a}$ has been proved by Van Hove [16] using Fourier transforms.

THEOREM 6.1. If e is in L_2 on G, there is a unique solution u of (2.1) in $\mathfrak{P}_2$ on G which vanishes on G^*. We have

$$d(u, G) \leqq l(e, G).$$

PROOF. If $\overline{a}_{ij}^{\alpha\beta} = \delta^{\alpha\beta}\delta_{ij}$, this follows since we may take the solution w_0 given by (2.2) and subtract off the harmonic function $H = w_0$ on G^* to obtain the desired solution w; remember $M = 1$ in that case. For the general $\overline{a}$ and any u in V_2 (i.e., in $\mathfrak{P}_2$ and vanishing on G^* with $\|u\|_0 = d(u, G)$) define T_0u as the solution w

in V_2 on G of

$$\int_{R^*} (\overline{a}_{ij}^{\alpha\beta} u^j_{x^\beta} - w^i_{x^\alpha})\, dx'_\alpha = 0$$

w exists in view of the preceding remark. If u is of class C''' on G and vanishes on and near G^*, we obtain

$$(T_0 u, u) = \int_G u^i_{x^\alpha} w^i_{x^\alpha}\, dx = \int_\Gamma u^i_{x^\alpha} w^i_{x^\alpha}\, dx = - \int_\Gamma u^i \Delta w^i\, dx$$

$$= - \int_\Gamma u^i \overline{a}_{ij}^{\alpha\beta} u^j_{x^\alpha x^\beta}\, dx = \int_G \overline{a}_{ij}^{\alpha\beta} u^i_{x^\alpha} u^j_{x^\beta}\, dx \geqq \|u\|_0^2$$

Γ being a smooth domain in G such that $u = 0$ on and near Γ^*. Since from above,

$$\|u\|_0^2 \leq (T_0 u, u) \leq \tfrac{1}{2}\|T_0 u\|_0^2 + \tfrac{1}{2}\|u\|_0^2$$

or

$$\|u\|_0^2 \leq \|Tu\|_0^2$$

we see that, in general, T_0 has an inverse of norm ≤ 1. Solving (2.1) is thus reduced to first solving

$$\int_{R^*} (w^i_{x^\alpha} + e^\alpha_i)\, dx'_\alpha = 0$$

for w in terms of e and then solving $T_0 u = w$ for u in terms of w.

THEOREM 6.2. Suppose that H^* is $\overline{a}$-harmonic and in $\mathfrak{P}_2$ on G and suppose H is harmonic on G and $H = H^*$ on G^*. Then

$$D_2(H^*, G) \leqq (1 + M_1^2) D_2(H, G)$$

If u is any function of class $\mathfrak{P}_2$ on G, there is a unique $\overline{a}$-harmonic function on G which coincides with u on G^*.

PROOF. Let $H^* = H + U$; then $U = 0$ on G^* and satisfies

(6.2) $$\int_{R^*} (\bar{a}_{ij}^{\alpha\beta} U_{x\beta}^j + \bar{a}_{ij}^{\alpha\beta} H_{x\beta}^j)\, dx'_\alpha = 0$$

From (6.1) and Theorem 6.1, we obtain

$$D_2(U, G) \leqq M_1^2 D_2(H, G)$$

The last statement follows by letting $H = u$ on G^* and solving (6.2) for U.

THEOREM 6.3. Suppose that H^* is $\bar{a}$-harmonic and in L_2 on G_a and is in $\mathfrak{P}_2$ on G_r and vanishes on σ_r for each $r < a$. Then H^* is analytic on σ_a also and we have

(6.3) $$|\nabla^n H^*(P)| \leqq C_9(\nu) d_0(H^*, G_a) \cdot n! (Q_1/d_P)^{n+\rho}$$

$$P \in G_a, \quad n = 0, 1, \ldots,$$

where Q_1 depends only on ν, N, and M_1.

PROOF. Let u be H^*, extended by reflection (i.e., (5.1)) to C_a. For each $r < a$, Let H_r be the harmonic function coinciding with u on C_r^*. By the principle of reflection, $H = 0$ on σ_r and so $H = H^*$ on G_r^*. Hence, from Theorem 6.2 and doubling, we see that u satisfies (5.11) with $K = Z_1 = (1 + M_1^2)$. Thus we obtain

(6.4) $$D_2(u, C_r) \leqq Z_2^2 L_2(u, C_a) \cdot (a - r)^{-2}, \quad 0 \leqq r < a, \; Z_2 = 2Z_1$$

Now, choose $r < a$, $0 < h_0 < a - r$, $r < r' < a - h_0$, $1 \leqq \alpha \leqq \nu - 1$, and for each h with $0 < h < h_0$, define

(6.5) $$u_{h\alpha}(x) = h^{-1}[u(x^\alpha + h, x'_\alpha) - u(x^\alpha, x'_\alpha)], \quad x \in C_{a-h}$$

Clearly each such $u_{h\alpha}$ has the properties of u so that

$$D_2(u_{h\alpha}, C_r) \leqq Z_2^2 L_2(u_{h\alpha}, C_{r'})(r' - r)^{-2}$$

(6.6) $$\sum_{\alpha=1}^{\nu-1} D_2(u_{h\alpha}, C_r) \leqq Z_2^2 D_2(u, C_{r'+h})(r' - r)^{-2}$$

$$\leqq Z_2^4 L_2(u, C_a) \cdot (a - r' - h)^{-2}(r' - r)^{-2}$$

Now as $h \longrightarrow 0$, $u_{h\alpha}(x) \longrightarrow u_{x\alpha}(x)$ for $x \in G_a$. Since the Dirichlet integral of $u_{h\alpha}$ remains bounded, a subsequence of $h \longrightarrow 0$ exists such that each $u_{h\alpha} \longrightarrow u_{x\alpha}$ in $\mathfrak{P}_2$ in each C_r. The lower-semicontinuity of D_2 allows us to let $h \longrightarrow 0$ in (6.6). Taking the minimum for $r < r' < a$, we obtain

$$\sum_{\alpha=1}^{\nu-1} D_2(u_\alpha, C_r) \leqq 2A_2^2, \quad A_2 = e^{-1}(eZ_2)^2 \cdot (2!) \cdot (a-r)^{-2} L_2(H^*, G_a)$$

Repeating this process and taking half for G_a, we obtain

$$\sum_{\alpha_1,\ldots,\alpha_p=1}^{\nu-1} D_2(u_{\alpha_1,\ldots\alpha_p}, G_r) \leqq A_{p+1}^2,$$

(6.7)

$$A_{p+1} = e^{-1}(eZ_2)^{p+1}(p+1)! \cdot (a-r)^{-p-1} \cdot L_2(H^*, G_a)$$

Now (6.7) shows that all the derivatives

$$H^*_{\alpha_1,\ldots,\alpha_{p+1}}, \quad H^*_{\nu,\alpha_1,\ldots,\alpha_p}$$

are in L_2 on each G_r $(r < a)$ with bounds (6.7). From the ellipticity, we may solve (1.1) for the $H^{*i}_{\nu\nu}$ obtaining

$$(6.8) \qquad H^{*i}_{\nu\nu} = \sum_{j=1}^{N} \sum_{\alpha=1}^{\nu} \sum_{\beta=1}^{\nu-1} A^{\alpha\beta}_{ij} H^{*j}_{\alpha\beta}, \quad \sum (A^{\alpha\beta}_{ij})^2 \leqq Z_3^2, \qquad x \in G_a$$

where Z_3 depends only on ν, N, and M_1. Using equations (6.8) and those obtained by differentiation, we can add in derivatives involving more derivatives with respect to x^ν to obtain

$$d_n(H^*, G_r) \leqq Z_4^n \cdot n! \cdot (a-r)^{-n} d_0(H^*, G_a)$$

(6.9)

$$Z_4 = Z_4(\nu, N, M_1)$$

To prove (6.3), let P be any point of G_a. Define

$$b = td_P/(2Z_4 + t), \quad 0 < t < 1$$

(6.10)

$$\text{so} \quad t = 2Z_4 b/(d_P - b) \quad \text{and} \quad 0 < b < d_P$$

From (6.9) and Lemma 5.7 (applied to components of $\nabla^n H^*$), we obtain

$$d_p[H^*, G(P, b)] \leqq d_0(H^*, G_a) \cdot p! Z_4^p \cdot (d_p - b)^{-p}$$

and

$$(6.11) \qquad |\nabla^n H^*(P)| \leqq Z_5 \left\{ \sum_{j=0}^{s} \frac{(2b)^j}{j!} d_{j+n}[u, G_b] + \frac{2(2b)^{s+1}}{s!} d_{s+n+1}[u, G_b] \right\}$$

$$G_b = G(P, b), \quad Z_5 = (2/\gamma_\nu)^{1/2} b^{-\rho}, \quad s \geqq [\rho]$$

From (6.10) and (6.11), we find that

$$(6.12) \qquad |\nabla^n H^*(P)| \leqq Z_5 d_0(H^*, G_a) Z_4^n (d_p - b)^{-n} \left[\sum_{j=0}^{s} \frac{(j+n)!}{j!} t^j + 2 \cdot \frac{(s+n+1)!}{s!} t^{s+1} \right]$$

Since $0 < t < 1$, we may let $s \longrightarrow \infty$ in (6.12). Using (6.10), we find that

$$(6.13) \qquad |\nabla^n H^*(P)| \leqq d_0(H^*, G_a) n! \left[\frac{2Z_4 + 1}{d_p} \right]^{n+\rho} \left[\frac{\sqrt{2/\gamma_\nu}}{2^n t^\rho (1-t)^{n+1}} \right]$$

It is easy to see that there is a number C_9 depending only on ν such that the minimum of the last bracket for $0 < t < 1$ is $\leqq C_9$ for each n. The analyticity on σ_a and (6.3) now follow.

THEOREM 6.4. Suppose H^* is $\bar{a}$-harmonic and in $\mathfrak{B}_2$ on $G(P, a)$, $P \in E_\nu^+ \cup \sigma$, and $H^* = 0$ on $C(P, a) \cap \sigma$. Then

(a) $d[H^*, G(P, r)] \leqq Q_2(\nu, N, M_1) \cdot d[H^*, G(P, a)](r/a)^\rho$, $0 \leqq r \leqq a$;

(b) if H is the harmonic function $= H^*$ on $G^*(P, a)$ and if l_r and l_r^* are the linear functions of Lemma 5.4 for H and H^* on $G(P, r)$, then

$$d[H^* - l_r^*, G(P, r)]$$

$$\leqq Q_3(\nu, N, M_1) \cdot d[H - l_a, G(P, a)] \cdot (r/a)^{\rho+1}, \quad 0 \leqq r \leqq a$$

PROOF. (a) is an immediate consequence of (1.10) and Theorem 6.3. To prove (b), let l^* be the linear function tangent to H^* at P. Then we note that $l^* - l_a$ is tangent to $H^* - l_a$ at P and also, using Theorem 6.2, that

$$(6.14) \qquad d[H^* - l_a, G(P, a)] \leqq (1 + M_1^2)^{1/2} d[H - l_a, G(P, a)]$$

Then, using (1.10) for $x_P \geqq a/2$ or Theorem 6.3 for $x_P \leqq a/2$, both applied to H_α^* $(\alpha = 1, \ldots, \nu)$, we obtain

$$(6.15) \qquad \begin{aligned} |\nabla H^*(P) - \nabla l_a(P)| &\leqq Z_1 a^{-\rho} d[H^* - l_a, G(P, a)] \\ &\leqq Z_2 a^{-\rho} d[H - l_a, G(P, a)] \end{aligned}$$

$$(6.16) \qquad \begin{aligned} &|\nabla H^*(x) - \nabla H^*(P)| \\ &\leqq Z_3 a^{-\rho-1} d[H^* - l^*, G(P, a)] \cdot |x - x_P|, \quad |x - x_P| \leqq a/4 \end{aligned}$$

Z_1, Z_2, Z_3 depending only on ν, N, and M_1. Since $\nabla l^* = \nabla H^*(P)$, we obtain

$$d[H^* - l^*, G(P, a)] \leqq Z_4 d[H - l_a, G(P, a)]$$

by means of (6.14) and (6.15).

If we use the fact that

$$d[H^* - l_r^*, G(P, r)] \leqq d[H^* - l^*, G(P, r)]$$

and (6.16), the result follows.

THEOREM 6.5. *Suppose u is the solution of (2.1) in V_2 on G(P, a) where e satisfies*

$$(6.17) \qquad l[e, G(P, r)] \leqq L(r/a)^\lambda, \quad 0 \leqq r \leqq a, \quad 0 \leqq \lambda < \rho$$

Then

$$d[u, G(P, r)] \leqq Q_4(\lambda, \nu, N, M_1) \cdot L(r/a)^{\lambda}$$

$$0 \leqq r \leqq a, \quad Q_4 \geqq 1$$

PROOF. This proof is similar to that of Theorem 2.2; we make the same definitions with obvious modifications. For each R with $0 < r \leqq R \leqq a$, write $u = U_R + H_R^*$ where $H_R^* = u$ on $G^*(P, R)$ and H_R^* is $\bar{a}$-harmonic on $G(P, R)$. Then $H_R^* = 0$ on $C(P, R) \cap \sigma$ so that

$$\begin{aligned} d[u, G(P, r)] &\leqq d[U_R, G(P, r)] + d[H_R^*, G(P, r)] \\ &\leqq L(R/a)^{\lambda}\phi(r/R) + Q_2 d[H_R^*, G(P, R)] \cdot (r/R)^{\rho} \\ &\leqq L(R/a)^{\lambda}\phi(r/R) + Z_1 d[u, G(P, R)] \cdot (r/R)^{\rho} \end{aligned}$$

$$Z_1 = Q_2(1 + M_1^2)$$

according to Theorems 6.4 and 6.2. Setting $s = r/a$, $t = R/a$, we obtain

$$\phi(s) \leqq t^{\lambda}\phi(s/t) + Z_1(s/t)^{\rho}\phi(t), \qquad 0 < s \leqq t \leqq 1$$

The result follows from Lemma 2.1.

LEMMA 6.2. Suppose ϕ is non-decreasing on $[0, 1]$ with $\phi(0) = 0$ and $\phi(1) \leqq M$ and suppose ϕ satisfies

$$\phi(s) \leqq K\phi(t) \cdot (s/t)^{\rho+1} + Lt^{\rho+\mu}(s/t)^{\rho-\epsilon} \tag{6.18}$$

$$0 < s \leqq t \leqq 1, \; 0 < \mu < 1, \; 0 \leqq \epsilon \leqq \rho$$

Then

$$\phi(s) \leqq S_1(\mu, \epsilon, \nu, K, L, M) \cdot s^{\rho+\mu}, \; 0 \leqq s \leqq 1 \tag{6.19}$$

PROOF. Let S_1 be any number such that

$$S_1 \geqq M, \; S_1 \geqq KM + L, \; S_1 \geqq \left[\frac{\lambda L}{(1-\lambda)KS_1}\right]^{-\lambda} \cdot \left[\frac{\lambda LM}{(1-\lambda)S_1} + L\right]$$

(6.20)

$$S_1 > \frac{\lambda L}{(1-\lambda)K}, \quad \lambda = \frac{\mu+\epsilon}{1+\epsilon}$$

For any such S_1, define $\sigma < 1$ by

$$\sigma^{1+\epsilon} = \frac{\lambda L}{(1-\lambda)KS_1} \tag{6.21}$$

Setting $t = 1$ in (6.18), we find that

$$s^{-\rho-\mu}\phi(s) \leqq KMs^{1-\mu} + Ls^{-\epsilon-\mu}, \qquad \sigma \leqq s \leqq 1$$

Since the maximum of this expression occurs at $s = \sigma$ or 1, we find that (6.19) holds for $\sigma \leqq s \leqq 1$. Now suppose $\sigma^2 \leqq s \leqq \sigma$. Then

$$\phi(s) \leqq KS_1 s^{\rho+1} t^{\mu-1} + Ls^{\rho-\epsilon} t^{\mu+\epsilon}, \quad \sigma \leqq t \leqq 1 \tag{6.22}$$

The minimum of the expression on the right occurs at $t = \sigma^{-1}s$. Substituting this value in we see, using (6.20), that (6.19) holds for $\sigma^2 \leqq s \leqq 1$. This may be repeated.

THEOREM 6.6. Suppose e is in L_2 on G_a and satisfies

$$l(e, G_a) \leqq L, \quad |e(\xi) - e(x)| \leqq Ld_x^{-\rho-\mu}|\xi - x|^{\mu}$$

(6.23)

$$|\xi - x| \leqq d_x/2; \; \xi, \; x \in G_a \cup \sigma_a, \qquad 0 < \mu < 1$$

Suppose u is the solution of (2.1) in V_2 on G_a. Then

$$|\nabla u(\xi) - \nabla u(x)| \leqq Q_5(\mu, \; \nu, \; N, \; M_1) \cdot Ld_x^{-\rho-\mu}|\xi - x|^{\mu}$$

$$|\xi - x| \leqq d_x/2; \; \xi, \; x \in G_a \cup \sigma_a$$

PROOF. We shall first show that

$$d[u - l_{P,r}, G(P, r)] \leqq Z_1(\mu, \; \nu, \; N, \; M) \cdot (r/d_P)^{\rho+\mu}$$

$$0 \leqq r \leqq d_P, \quad P \in G_a \cup \sigma_a$$

where $l_{P,r}$ is the linear function of Lemma 5.4 for u and H_{Pr} on $G(P, r)$ where H_{Pr} is harmonic on $G(P, r)$ and $= u$ on $G^*(P, r)$. The result follows from Lemmas 5.5 and 5.6.

We note first that (6.23) implies that

$$l[e - e(P), G(P, r)] \leqq Z_2(\mu, \nu) \cdot L(r/d_P)^{\rho+\mu} \tag{6.24}$$
$$0 \leqq r \leqq d_P, \ P \in G_a \cup \sigma_a$$

For a fixed P and s, let

$$\phi(s) = \underset{0 \leqq r' \leqq sd_P}{\text{l.u.b.}} L^{-1} d[u - l_{P,r'}, G(P, r')], \ 0 \leqq s \leqq 1$$

Then ϕ is non-decreasing on $[0, 1]$ with $\phi(0) = 0$, $\phi(1) \leqq 1$ by Theorem 6.1. Now, choose $0 \leqq r' \leqq r \leqq R \leqq d_P$, $r > 0$, and write $u = U_{P,R} + H^*_{P,R}$ where $H^*_{P,R} = u$ on $G^*(P, R)$ and $H^*_{P,R}$ is $\bar{a}$-harmonic on $G(P, R)$. Since $U_{P,R}$ is not altered when a constant is subtracted from e, we see that $U_{P,R}$ is the solution of (2.1) in V_2 on $G(P, R)$ with e replaced by $e - e(P)$. Using (6.24) on $G(P, R)$ and Theorems 6.4 and 6.5, we obtain

$$d[u - l_{P,r'}, G(P, r')] \leqq d[u - l^*_{P,Rr'}, G(P, r')] \leqq d[u - l^*_{PRr}, G(P, r)]$$
$$\leqq d[U_{PR}, G(P, r)] + d[H^*_{PR} - l^*_{PRr}, G(P, r)]$$
$$\leqq Z_3(\epsilon, \ldots)L(R/d_P)^{\rho+\mu}(r/R)^{\rho-\epsilon} + Q_3 d[H_{PR} - l_{PR}, G(P, R)] \cdot (r/R)^{\rho+1} \tag{6.25}$$
$$\leqq Z_3(\epsilon, \ldots)L(R/d_P)^{\rho+\mu}(r/R)^{\rho-\epsilon} + Q_3 L\phi(R/d_P) \cdot (r/R)^{\rho+1}$$
$$0 < \epsilon \leqq \rho, \ Z_3(\epsilon, \ldots) = Z_2 \cdot Q_4(\rho - \epsilon, \ldots)$$

where l^*_{PRr} is the function of Lemma 5.4 for H^*_{PR} on $G(P, r)$. Setting $s = r/d_P$ and $t = R/d_P$ in (6.25), we obtain

$$\phi(s) \leqq Q_3\phi(t) \cdot (s/t)^{\rho+1} + Z_3(\epsilon, \ldots)t^{\rho+\mu}(s/t)^{\rho-\epsilon}$$
$$0 < s \leqq t \leqq 1$$

The result follows from Lemma 6.2 after we take the minimum of the result for all the ϵ allowed.

THEOREM 6.7. Suppose u is the solution of (2.1) in V_2 on G_a where e satisfies (6.17) on all $G(P, d_P) \subset G_a$. Then

$$d[u, G(P, r)] \leqq Q_6 L(r/d_P)^{\lambda}, \quad 0 \leqq r \leqq d_P, \quad P \in G_a, \quad Q_6 = Q_4 + 2Q_2$$

PROOF. On each $G(P, d_P)$, write $u = U_P + H_P^*$ where U_P is the solution of (2.1) in V_2 on $G(P, d_P)$. Then, from Theorems 6.5, 6.4, and 6.1, we obtain

$$d[U_P, G(P, r)] \leqq Q_4 L(r/d_P)^{\lambda}, \quad d[U_P, G(P, d_P)] \leqq L$$

$$d[u, G(P, d_P)] \leqq L, \quad d[H_P^*, G(P, d_P)] \leqq 2L$$

$$d[H_P^*, G(P, r)] \leqq 2Q_2 L(r/d_P)^{\rho}$$

The result follows by combining the first and last of these.

For the remainder of this section, we assume that the coefficients $a(x)$ to $d(x)$ satisfy the fundamental hypotheses of §§3 and 4 on some $\overline{G}_{a_0}$; any G_a which we discuss will have $a \leqq a_0$.

DEFINITION 6.1. For functions u in V_2 on G_a, we define various norms as follows:

$$\|u\|_0 = d(u, G_a)$$

$$\|u\|_{0\lambda} = \underset{P \in G_a}{\text{l.u.b.}} \left\{ \underset{0<r\leqq d_P}{\text{l.u.b.}} (r/a)^{-\lambda} d[u, G(P, r)] \right\} \geqq \|u\|_0$$

$\|u\|_{1\mu}$ = g.l.b. of numbers L such that

$\|u\|_0 \leqq L$ and

$$|\nabla u(\xi) - \nabla u(x)| \leqq L d_x^{-\rho-\mu} |\xi - x|^{\mu}, \quad |\xi - x| \leqq d_x/2, \quad x, \xi \in G_a \cup \sigma_a$$

the corresponding spaces of functions with finite norms will be denoted by V_2, $V_{2\lambda}$, and $C_{1\mu}$.

DEFINITION 6.2. For all u in V_2 on G_a, we define $T_3 u$ as the solution U of (2.1) in V_2 on G_a with e replaced by $(a - \overline{a}) \cdot \nabla u + b \cdot u - \nabla v$, v being the C_a-potential of f where

$$f(x) = \begin{cases} c(x) \cdot \nabla u(x) + d(x) \cdot u(x), & x \in G_a \cup \sigma_a \\ f(-x^\nu, x'_\nu) & , x \in C_a - G_a \end{cases} \tag{6.26}$$

THEOREM 6.8. The transformation T_3 is an operator from V_2 to V_2 and from $V_{2\lambda}$ to $V_{2\lambda}$ and we have

$$\|T_3\|_0 \leqq k + (M_2 + \nu^{1/2}C_2M_3)a + \nu^{1/2}C_2^2a^2 \tag{6.27}$$

$$\|T_3\|_{0\lambda} \leqq Q_6[k + (C_{10}M_2 + C_{11}M_3)a + C_{12}M_4a^2], \quad C_n = C_n(\lambda, \nu) \tag{6.28}$$

If, in addition, $\bar{a} = \bar{a}(0)$ and a and b satisfy (3.22) on $\bar{G}_{a_0}$, then T_3 is an operator from $C_{1\mu}$ to $C_{1\mu}$ and

$$\|T_3\|_{1\mu}$$

$$\leqq Q_5[C_{13}A_0a^\mu + C_{14}M_2a + C_{15}M_3a + C_{16}M_4a^2 + C_{17}B_0a^{1+\mu}] \tag{6.29}$$

$$C_n = C_n(\mu, \nu)$$

PROOF. We see that v satisfies

$$v(-x^\nu, x'_\nu) = v(x^\nu, x'_\nu), \quad x \in C_a \tag{6.30}$$

If ϕ satisfies either (5.1) or (6.30), we have

$$\begin{aligned} d_k(\phi, C_a) &= 2^{1/2}d_k(\phi, G_a) \\ d_k[\phi, G(P, r)] &\leqq d_k[\phi, C(P, r)] \\ &\leqq 2^{1/2}d_k[\phi, G(P, r)] \end{aligned} \tag{6.31}$$

So, using (6.31) and Lemmas 3.1 to 3.4 and 3.7 and arguments like those in the proofs of Theorem 6.7 and the last part of Lemma 3.7 in order to get bounds on terms like $l[u, C(P, r)]$, $d_2[v, C(P, r)]$, $d_1[v, C(P, r)]$, etc., we obtain the results in much the same way that they were obtained in Theorems 3.1 and 3.2. It is to be remembered that $M = 1$ and bounds are independent of N if $\bar{a}_{ij} = \delta^{\alpha\beta}\delta_{ij}$.

§7. REGULARITY PROPERTIES OF THE SOLUTIONS OF (A) ON SMOOTH BOUNDED CLOSED DOMAINS

We assume that a, b, c, and d satisfy their usual hypotheses on a bounded closed domain $\overline{G}$. For the first part of this section we assume $\overline{G} = \overline{G}_{a_0}$.

THEOREM 7.1. Suppose u is in L_2 on G_a, is in $\mathfrak{P}_2$ on G_r and vanishes on σ_r for each $r < a$, and satisfies (A) on G_a. Then there are numbers $d_6 > 0$, Q_7, Q_8, and Q_9, depending only on ν, N, $M_1, \ldots, M_4$, and the modulus of continuity of a(x) such that

$$D_2(u, G_r)$$

$$\leqq Q_7(a - r)^{-2}L_2(u, G_a) + (a - r)^{-1}[Q_8 a L_2(e, G_a) + Q_9 a^3 L_2(f, G_a)]$$

provided $a \leqq d_6$ and e and f are in L_2 on G_a.

PROOF. Choose $d_6 > 0$ but so small that the right side of (6.27) $\leqq 1/2$, with $M_2 = M_4 = 0$, for all $a \leqq d_6$, $k = k(a)$. For each $r < a$, let $u = U + H$ where H is harmonic with $H = u$ on G_r^*; then $U = H = 0$ on σ_r, $U = 0$ on G_r^*. Let T_3 be the transformation of Theorem 6.8 with $b = d = 0$. Let V be the C_r-potential of $F = c \cdot \nabla H + d \cdot u + f$ on G_r, F satisfying $F(-x^\nu, x_\nu') = F(x)$ on C_r, and let W be the solution of (2.1) in V_2 on G_r with e replaced by $E = a \cdot \nabla H + b \cdot u + e - \nabla v$. Using Lemma 3.7 and Theorem 6.1, we obtain

$$d(W, G_r) \leqq M_1 d(H, G_r) + M_2 l(u, G_r) + l(e, G_R)$$

$$+ \nu^{1/2} C_2 r[M_3 d(H, C_r) + M_4 l(u, C_r) + l(f, C_r)]$$

Since $\|T_3\|_0 \leqq 1/2$ and $U - T_3 U = W$, we obtain

$$D_2(u, C_r)$$

$$\leqq [1 + 16Z_1^2]D_2(H, C_r) + 16Z_2^2 L_2(u, C_r) + 16L(e, C_r)$$

$$+ 16r^2 Z_3^2 L(f, C_r)$$

The result follows from Theorem 5.1.

THEOREM 7.2. Suppose u is of class $\mathfrak{B}_2$, satisfies (5.1) on C_a and satisfies (A) on G_a. Suppose e and f satisfy

$$l[e, G(P, r)] \leq L(r/d_P)^{\lambda}, \quad l[f, G(P, r)] \leq L(r/d_P)^{\lambda} \tag{7.1}$$

$$0 \leq r \leq d_P, \quad P \in G_a, \qquad 0 \leq \lambda < \rho$$

There are numbers $d_7 > 0$ and Q_{10} and Q_{11} depending only on λ and the quantities in Theorem 7.1 such that

$$d[u, C(P, r)] \leq [Q_{10} d(u, C_a) + Q_{11} L](r/d_P)^{\lambda}, \quad 0 \leq r \leq d_P$$

$$P \in C_a, \quad a \leq d_7$$

If $\lambda = \rho - 1 + \mu$, $0 < \mu < 1$, then u is of class C^0_μ on C_a and satisfies

$$\begin{aligned} &|u(\xi) - u(x)| \\ &\leq \mu^{-1} C_1 [Q_{10} d(u, C_a) + Q_{11} L] d_x^{1-\rho-\mu} |\xi - x|^{\mu} \end{aligned} \tag{7.2}$$

$$|\xi - x| \leq d_x/2, \quad x \in C_a$$

PROOF. Choose d_7 so small that the right side of (6.28) $\leq 1/2$ for $a \leq d_7$. Let T_3 be the operator of Theorem 6.8 on G_a. Let $u = U + H$ on C_a, H harmonic on C_a, $H = u$ on C_a^*; then $H = u = U = 0$ on σ_a, $U = 0$ on G_a^*. Let V be the C_a-potential of $F = c \cdot \nabla H + d \cdot H + f$, extended to C_a as usual; let W be the solution in V_2 on G_a of (2.1) with e replaced by $E = a \cdot \nabla H + b \cdot H + e - \nabla V$ and let W be extended to C_a by (5. 1). Since $U - T_3 U = W$ and $\|T_3\|_{0\lambda} \leq 1/2$, the result follows easily from Theorem 6.7, Lemma 3.1, etc.

THEOREM 7.3. Suppose u is of class $\mathfrak{B}_2$ and satisfies (A) on G_a and vanishes on σ_a. Suppose e and f satisfy (4.21) on G_a, (with $C(x, r)$ replaced by $G(x, r)$, δ_x by d_x), and suppose a and b satisfy (3.22) on $\overline{G}_{a_0}$. Then u is of class C'_μ on $G_a \cup \sigma_a$ and there are numbers $d_8 > 0$, Q_{12} and Q_{13}, depending only on the previous quantities, A_0 and B_0, such that

$$|\nabla u(\xi) - \nabla u(x)| \leqq [Q_{12} d(u, G_a) + Q_{13} L] d_x^{-\rho-\mu} |\xi - x|^{\mu}$$

$$|\xi - x| \leqq d_x/2, \quad \xi, x \in G_a \cup \sigma_a$$

provided $a \leqq d_8$.

The proof is like that of Theorem 7.3 and uses Theorem 6.6, Lemma 3.7, etc.

THEOREM 7.4. Suppose u satisfies the hypotheses of Theorem 7.1 on G_a, suppose $a \leqq d_6$ and d_8 and suppose a, b, and e are of class $C_\mu^{(n+1)}$ on $\overline{G}_a$ and c, d, and f are of class $C_\mu^{(n)}$ on $\overline{G}_a$, $n \geqq 0$, $0 < \mu < 1$. Then u is of class $C_\mu^{(n+2)}$ on each $\overline{G}_r$ with $r < a$, the bounds for $\nabla^k u$ and the Hölder coefficient for $\nabla^{n+2} u$ on $\overline{G}_r$ depending only on a, r, k, n, ν, N, $l(u, G_a)$, and on the bounds and Hölder constants for $a, \ldots, f$ and for their derivatives.

PROOF. We first prove this for $n = 0$. Choose $r < r_1 < r_2 < \ldots < r_5 < a$, all equally spaced. By Theorem 7.1, $d_1(u, G_{r_5}) \leqq Z_1$ and then by Theorem 7.3, u is of class C_μ' on $\overline{G}_{r_4}$ with corresponding bounds, etc. We let v be the C_{r_4} - potential of $c \cdot \nabla u + d \cdot u + f$, extended as usual across σ_{r_4}. Then v is of class C_μ^2 on $\overline{C}_{r_3}$ with proper bounds and u satisfies (2.1) on G_{r_3} with e replaced by $e + b \cdot u - \nabla v$, of class C_μ' on $\overline{G}_{r_3}$ and $u = 0$ on σ_{r_3}. For $0 < |h| < r_3 - r_2$, we apply the device of Lichtenstein to the resulting equation of the form (2.1), using only unit vectors ζ orthogonal to the x^ν direction. Application of Theorems 7.1 and 7.3 on $\overline{G}_r$ allows us to conclude that each u_{x^α}, $\alpha = 1, \ldots, \nu - 1$, is of class C_μ' on $\overline{G}_r$ and vanishes on σ_r. This shows that all the derivatives $u_{\alpha\beta}$ and $u_{\nu\alpha}$ are in C_μ^0 on G_r, $\alpha, \beta \leqq \nu - 1$. Since equations (D) hold on the <u>interior</u> of G_r, we may solve for the $u_{\nu\nu}$ and obtain the result.

For $n > 0$, we repeat the first part of the process finding that all the derivatives $u_{\alpha_1, \ldots, \alpha_{n+2}}$ and $u_{\nu, \alpha_1, \ldots, \alpha_{n+1}}$ are in C_μ^0. We then use equations (D) and their derivatives to obtain the result.

DEFINITION 7.1. A bounded domain G is said to be of class $C^{(n)}$ ($C_\mu^{(n)}$, $0 < \mu < 1$) if the boundary can be covered by a finite number of sets, each of which is open on $\overline{G}$ and is the image of $G_{a_0} \cup \sigma_{a_0}$ $(a_0 > 0)$ under a 1 - 1 transformation which is of class $C^{(n)}$ $(C_\mu^{(n)})$ together with its inverse.

We now summarize our results.

THEOREM 7.5. Suppose G is a bounded domain of class C', u^* is of class $\mathfrak{P}_2$ on G, e and f are in L_2 on G, and the coefficients $a, \ldots, d$ satisfy their usual conditions on $\overline{G}$. Suppose u is of class $\mathfrak{P}_2$ on G and satisfies (A) on G. Then

$$d_1(u, G) \leqq L_1 l(u, G) + L_2 l(e, G) + L_3 l(f, G) + L_4 d_1(u^*, G)$$

where $L_1, \ldots, L_4$ depend only on ν, N, $M_1, \ldots, M_4$, G, and the modulus of continuity of $a(x)$ on $\overline{G}$.

If, also, u^*, e, and f satisfy the following:

$$l[\phi, G \cap C(P, r)] \leqq Lr^{\rho-1+\mu}, \quad r > 0, \quad 0 < \mu < 1$$

where ∇u^*, e, or f may be inserted for ϕ; all $P \in \overline{G}$, then u^* and u are μ-Hölder continuous on $\overline{G}$, the bound for u and its Hölder constant depending only on the previous quantities, on μ and on L.

If, also, G is of class C_μ', a, b, and e are of class $C_\mu^{(o)}$ on $\overline{G}$, and the boundary values of u^* are of class C_μ' along G^*, then u is of class C_μ' on $\overline{G}$, the bound and Hölder coefficient for ∇u depending only on the previous quantities and on the Hölder coefficients for a, b, and e on $\overline{G}$ and on that of the tangential gradient of u^*.

If, also, G is of class $C_\mu^{(n+2)}$, $n \geqq 0$; if the boundary values for u^* are of class $C_\mu^{(n+2)}$ along G^*, if a, b and e are of class $C_\mu^{(n+1)}$ on $\overline{G}$, and if c, d and f are of class $C_\mu^{(n)}$ on $\overline{G}$, then u is of class $C_\mu^{(n+2)}$ on $\overline{G}$, the bounds for $\nabla^k u$ and the Hölder constant for $\nabla^{n+2} u$ depending only on corresponding quantities for u^* along G^* and on k, n, ν, N, and on the bounds and Hölder constants for a, ..., f and for their derivatives on $\overline{G}$.

Finally, if the hypotheses above for a, ..., f, u, and u^* hold on some neighborhood η of $\overline{G}$, then the corresponding conclusions hold on smaller neighborhoods; in this last statement, it is not necessary for the whole of G^* to be regular.

PROOF. We may write $u = u^* + U$, absorbing terms involving u^* into e and f; in the third and fourth paragraphs, we may replace u^* by the harmonic function coinciding with it on G^*. So we may assume $u^* = 0$. A simple argument shows that equations (A) are transformed into new equations (A), still elliptic, by changes of variable x. By beginning by mappings of fairly large neighborhoods of G^* onto $G_{a_0} \cup \sigma_{a_0}$ and then taking smaller parts $\overline{G(P, a)}$, $P \in \sigma$, etc., we reduce the above to Theorems 7.1 to 7.4. A finite number of these smaller neighborhoods still covers a neighborhood of G^* and the results follow from those of §4.

§8. THE EXISTENCE THEORY FOR EQUATIONS (A), MODIFIED, ON AN ARBITRARY BOUNDED DOMAIN G

In this section, we discuss the system (A) modified by replacing d by $d_{ij} + \lambda\delta_{ij}$, λ being a parameter and δ_{ij} being the Kronecker delta. Since we have already discussed regularity properties, we merely make our minimum assumptions about the coefficients.

LEMMA 8.1. (Gårding's lemma). Suppose the tensor a(x) is continuous on $\overline{G}$ (G bounded) and satisfied (6.1) there. Then for each $\epsilon > 0$, there is a constant λ_1 depending only on G, ϵ, M_1, and the modulus of continuity of a(x) such that

$$I_a(u, G) \equiv \int_G a_{ij}^{\alpha\beta} u_{x^\alpha}^i u_{x^\beta}^j \, dx \tag{8.1}$$
$$\geqq (1 - \epsilon) d_1^2(u, G) - \lambda_1 d_0^2(u, G)$$

for all u in V_2 on G.

PROOF. If a(x) is a constant tensor satisfying (6.1), we have seen in Lemma 6.1 that

$$I_a(u, \Gamma) \geqq d_1^2(u, \Gamma)$$

for all u in V_2 on any bounded open set Γ.

In general, we may choose a finite number of non-overlapping hypercubes $\bar{r}_p$ $(p = 1, \ldots, P)$ which cover $\bar{G}$, each of which contains a point x_p of $\bar{G}$ in its interior and small enough so that

$$|a(x) - a(x_p)| < \epsilon/2, \; x \in \bar{G} \cap R_p \tag{8.2}$$

where R_p is the concentric hypercube to $\bar{r}_p$ of twice the diameter. It is easy to see how to construct a sequence $\{\eta_p(x)\}$, each term of class C' everywhere and 0 outside R_p, such that

$$\sum_{p=1}^{P} \eta_p(x) = 1, \qquad x \in \bigcup_{p=1}^{P} \bar{r}_p \tag{8.3}$$

Then, making the replacement (8.3) for 1 on G, we obtain

$$\begin{aligned} I_a(u, G) &= \sum_{p=1}^{P} I_a(\eta_p u, G_p) - J(u, G) \\ &\geqq (1 - \epsilon/2) \sum_{p=1}^{P} d_1^2(\eta_p u, G_p) - J(u, G) \end{aligned} \tag{8.4}$$

$$G_p = G \cap R_p$$

using (8.2), where

$$\begin{aligned} J(u, G) = \int_G \Bigg\{ &\left[\sum_{p=1}^{P} (a_{ij}^{\alpha\beta} + a_{ji}^{\beta\alpha}) \eta_p \eta_{px^\beta} \right] u^i_{x^\alpha} u^j \\ &+ \left[\sum_{p=1}^{P} a_{ij}^{\alpha\beta} \eta_{px^\alpha} \eta_{px^\beta} \right] u^i u^j \Bigg\} dx \end{aligned} \tag{8.5}$$

Repeating the argument in reverse with $I_a(u, G)$ replaced by $D_2(u, G)$ we obtain

(8.6) $I_a(u, G) \geqq (1 - \epsilon/2)d_1^2(u, G) - J(u, G) + (1 - \epsilon/2)K(u, G)$

$$K(u, G) = \int_G \sum_{p=1}^{P} \left[2\eta_p \eta_{px^\alpha} u^i_{x^\alpha} u^i + |\nabla \eta_p|^2 \cdot |u|^2\right] dx$$

From the forms of $J(u, G)$ and $K(u, G)$, we see that there are positive constants J and K which depend only on M_1, G, and the modulus of continuity of $a(x)$ such that

(8.7)
$$\begin{aligned} I_a(u, G) &\geqq (1 - \epsilon/2)d_1^2(u, G) - Jd_0(u, G)d_1(u, G) - Kd_0^2(u, G) \\ &\geqq (1 - \epsilon)d_1^2(u, G) - (K + J^2/\epsilon^2)d_0^2(u, G) \end{aligned}$$

from which the lemma follows.

DEFINITION 8.1. We define the operators T and T_2 on V_2 on G as follows: For each u in V_2, let $T_2 u$ be the solution W in V_2 of

(8.8) $$\int_{R^*} W^i_{x^\alpha}\, dx'_\alpha = - \int_R u^i\, dx$$

and let Tu be the solution v in V_2 on G of

(8.9) $$\int_{R^*} (a^{\alpha\beta}_{ij} u^j_{x^\alpha} + b^\alpha_{ij} u^j - V^i_{x^\alpha} - v^i_{x^\alpha})\, dx'_\alpha = 0$$

where V is the G-potential of $c \cdot \nabla u + d \cdot u$; clearly W is the G-potential of $(- u)$ minus the harmonic function coinciding with it on G^*.

That T and T_2 are operators follows from Theorem 6.1 and Lemma 3.6.

LEMMA 8.2. If $v = Tu$ and $W = T_2 u$, we have

(i)
$$\begin{aligned} &\int_{R^*} (a^{\alpha\beta}_{ij} u^j_{x^\beta} + b^\alpha_{ij} u^j - v^i_{x^\alpha})\, dx'_\alpha \\ &= \int_R (c^\alpha_{ij} u^j_{x^\alpha} + d_{ij} u^j)\, dx \end{aligned}$$

(ii) $$(u, Tu) = \int_G u^i_{x^\alpha} v^i_{x^\alpha}\, dx = I_a(u, G) + J(u, G)$$

$$J(u, G) = \int_G [u^i_{x^\alpha}(b^\alpha_{ij} + c^\alpha_{ji})u^j + d_{ij}u^i u^j]\, dx$$

(iii) $$(u, T_2 u) = \int_G u^i_{x^\alpha} W^i_{x^\alpha}\, dx = \int_G |u|^2\, dx$$

(iv) T_2 is completely continuous.

PROOF. (i) follows from the definition of V and Lemma 3.6. (iv) follows from Lemma 3.6, Theorem 6.1, and the fact that $u_n \rightrightarrows u$ in L_2 on G whenever $u_n \longrightarrow u$ in V_2 on G. From Haar's lemma ([12], Chap. V, §§1, 2) we have

$$\int_{R^*} A^\alpha_i\, dx'_\alpha = \int_R B_i\, dx$$

if and only if

$$\int_G (A^\alpha_i z^i_{x^\alpha} + B_i z^i)\, dx = 0$$

for every z in V_2 on G, A and B being in L_2 on G. Then (ii) and (iii) follow by applying this to (i) and to (8.8) with $z = u$.

Now, suppose w is the solution in V_2 on G of

$$\int_{R^*} (w^i_{x^\alpha} + e^\alpha_i)\, dx'_\alpha = \int_R f_i\, dx \quad \text{or} \quad \int_{R^*} (w^i_{x^\alpha} + e^\alpha_i - z^i_{x^\alpha})\, dx'_\alpha = 0$$

z being the G-potential of f, and let $W = T_2 u$. Then u is a solution of our modified (A) in V_2 on G if and only if u satisfies

$$\int_{R^*} (a^{\alpha\beta}_{ij} u^j_{x^\beta} + b^\alpha_{ij} u^j - w^i_{x^\alpha})\, dx' = \int_R (c^\alpha_{ij} u^j_{x^\alpha} + d_{ij} u^j)\, dx$$

$$+ \lambda \int_R u^i dx$$

or

$$\int_{R*} a_{ij}^{\alpha\beta} u_{x^\beta}^j + b_{ij}^{\alpha} u^j - (w_{x^\alpha}^i - \lambda W_{x^\alpha}^i)\, dx'$$

$$= \int_R (c_{ij}^{\alpha} u_{x^\alpha}^j + d_{ij} u^j)\, dx$$

Comparing with Lemma 8.4 (i) and the definitions of T and T_2, we see that this is equivalent to the statement

$$(8.10) \qquad Tu + \lambda T_2 u = w$$

THEOREM 8.1. The modified system (A) has a unique solution u in V_2 on G for any given e and f in L_2 unless λ belongs to an isolated set of characteristic values which set is bounded above.

PROOF. From Lemma 8.2, it follows that

$$(u, Tu + \lambda T_2 u) = I_a(u, G) + J(u, G) + \lambda d_o^2(u, G)$$

As in the proof of Lemma 8.1, we may obtain

$$J(u, G) \geqq - J d_0(u, G) d_1(u, G) - K d_o^2(u, G)$$

Using Lemma 8.1 with $\epsilon = 1/4$, we obtain

$$(u, Tu + \lambda T_2 u)$$

$$\geqq \tfrac{3}{4} d_1^2(u, G) - (\lambda_1 + K) d_o^2(u, G) - J d_0(u, G) d_1(u, G) + \lambda d_o^2(u, G)$$

$$\geqq \tfrac{1}{2}\|u\|^2 + (\lambda - \lambda_0) d_o^2(u, G)$$

$$\lambda_0 = \lambda_1 + K + J^2$$

Hence if $\lambda \geqq \lambda_0$, $T + \lambda T_2$ has an inverse with norm $\leqq 2$ so that (8.10) has a unique solution for u in terms of w if $\lambda \geqq \lambda_0$. If we define $T_0 = T + \lambda_0 T_2$, then T_0 has an inverse of norm $\leqq 2$. Applying T_0^{-1} to both sides of (8.10), we obtain

$$(8.11) \qquad [I + (\lambda - \lambda_0) T_0^{-1} T_2] u = T_0^{-1} w$$

Since $T_0^{-1}T_2$ is completely continuous the result follows from the Riesz theory (see [1], Chapter X, for instance).

THEOREM 8.2. If λ is not a characteristic value (as defined above) then the modified system (A) has a unique solution u which coincides on G^* with any given function u^* of class $\mathfrak{P}_2$ on G. If λ is characteristic, the homogeneous system $(e = f = 0)$ has a non-zero solution which vanishes on G^*, the linear manifold of these being finite dimensional.

PROOF. To prove the first statement, write $u = U + u^*$. Then $U = 0$ on G^* and satisfies (A) modified with e and f replaced by $a \cdot \nabla u^* + b \cdot u^* + e$ and $c \cdot \nabla u^* + d \cdot u^* + f$, both still in L_2. Since U exists and is unique, the result follows. The last statement follows since then (8.11) has a finite dimensional manifold of solutions when $w = 0$.

§9. APPLICATIONS TO NON-LINEAR SYSTEMS AND TO THE CALCULUS OF VARIATIONS

We begin by considering the general elliptic system

$$\phi_i(x, z, \nabla z, \nabla^2 z) = 0, \qquad i = 1, \ldots, N \tag{9.1}$$

where the ϕ_i are assumed to be of class C'. The ellipticity condition is

$$\text{determinant of } \left\| A^{\alpha\beta}_{kj}\lambda_\alpha\lambda_\beta \right\| \neq 0, \quad \lambda \neq 0 \tag{9.2}$$

where

$$A^{\alpha\alpha}_{ij} = \partial\phi_i/\partial r^j_{\alpha\alpha}$$

$$A^{\alpha\beta}_{ij} = \begin{cases} \frac{1}{2}\partial\phi_i/\partial r^j_{\alpha\beta}, & \alpha < \beta \\ \\ \frac{1}{2}\partial\phi_i/\partial r^j_{\beta\alpha}, & \beta < \alpha \end{cases}, \quad (\alpha \text{ not summed}) \tag{9.3}$$

the second definition being made to allow symmetric sums in (α, β) (of course the distinct arguments are $r^j_{\alpha\beta}$ with $\alpha \leq \beta$).

THEOREM 9.1. Suppose z is of class C'' and satisfies (9.1) on a domain including $\overline{G}$ (bounded). Suppose the ϕ_i are of class C' and satisfy (9.2) on $\overline{\mathcal{G}}$, $\mathcal{G}$ being a domain which contains the compact set of all $[x, z(x), \nabla z(x), \nabla^2 z(x)]$ for x on $\overline{G}$. Then $\nabla^2 z$ is of class $\mathfrak{P}_2$ and of class $C_\mu^{(0)}$ on each domain interior to G for any μ, $(0 < \mu < 1)$. If, also, the ϕ_i are of class $C_\mu^{(n)}$ on $\overline{\mathcal{G}}$, $(n \geq 1, 0 < \mu < 1)$ then z is of class $C_\mu^{(n+2)}$ on G.

PROOF. Choose $D \subset \overline{D} \subset D_1 \subset \overline{D}_1 \subset D_2 \subset \overline{D}_2 \subset G$. For each unit vector ζ and each h, $(0 < |h| < h_o =$ the distance of $\overline{D}_2$ from $G^*)$ we define

$$(9.4) \qquad u(x) = h^{-1}[z(x + h\zeta) - z(x)]$$

$$(9.5) \qquad a(x) = \int_0^1 A\left\{x + th\zeta,\ Z(x),\ \nabla Z(x),\ \nabla^2 z(x)\right\} dt$$

$$Z(x) = z(x) + t[z(x + h) - z(x)]$$

$$(9.6) \quad f(x) = -\nabla u(x) \cdot \int_0^1 \phi_p(*)\, dt - u(x) \cdot \int_0^1 \phi_z(*)\, dt - \zeta \cdot \int_0^1 \phi_x(*)\, dt$$

Applying the device of Lichtenstein to (9.1) for these h, we see that

$$(9.7) \qquad a_{ij}^{\alpha\beta}(x) u^j_{x^\alpha x^\beta}(x) = f_i(x),\ x \in D_2$$

If we restrict $|h| \leq h_1 < h_o$, if necessary, we see from (9.5) and (9.6) that $a(x)$ and $f(x)$ are equicontinuous and uniformly bounded and that $a(x)$ satisfies (1.6) for some M independent of h. Applying Theorem 4.7, we see that we may let $h \longrightarrow 0$ to obtain the first result.

If, also, ϕ is of class C'_μ on $\overline{\mathcal{G}}$, we use the result above to show that $a(x)$ and $f(x)$ are in $C^{(0)}_{\mu-\epsilon}$ on any $\overline{D}$ independently of h from which we conclude that $\nabla^2 u$ is equi-$(\mu - \epsilon)$-Hölder continuous, then that $a(x)$ and $f(x)$ are equi-μ-Hölder continuous and hence that $\nabla^2 u$ is also, all independently of h. The second result follows for $n = 1$ and the higher regularity results are obtained by repetition.

DEFINITION 9.1. A function z of class C' on $\overline{G}$ (bounded) is said to furnish a stationary value to the integral

(9.8) $$I(z, G) = \int_G f[x, z(x), \nabla z(x)]\, dx$$

if and only if the first variation

(9.9) $$J(z, w; G) = \int_G [f_p(*) \cdot \nabla w(x) + f_z(*) \cdot w(x)]\, dx = 0$$

$$(*) = [x, z(x), \nabla z(x)]$$

for every w of class C' on $\bar{G}$ which vanishes on and near G^*.

It follows from Haar's lemma ([12], Chapter V, §§1,2) that (9.9) is equivalent to the condition that z satisfies Haar's equations.

(9.10) $$\int_{R^*} f_{p^i_\alpha}(*)\, dx'_\alpha = \int_R f_{z^i}(*)\, dx$$

DEFINITION 9.2. Suppose $z(x)$ is of class C' on $\bar{G}$. The function $f(x, z, p)$ is said to be the integrand of a regular variational problem near $z(x)$ if and only if f and f_p are of class C' and f satisfies

(9.11) $$f_{p^i_\alpha p^j_\beta}(x, z, p)\lambda_\alpha\lambda_\beta\xi^i\xi^j > 0, \quad |\lambda| \neq 0, \quad \xi \neq 0$$

for all $(x, z, p) \in \bar{\mathcal{G}}$, $\mathcal{G}$ being a bounded domain of the form

$$\mathcal{G} : x \in G, \; |p - p(x)| < h, \; |z - z(x)| < h, \; h > 0$$

In what follows, we shall assume that f satisfies

(9.12) $$\begin{gathered} f_{p^i_\alpha p^j_\beta}(x, z, p)\lambda_\alpha\lambda_\beta\xi^i\xi^j \geqq |\lambda|^2 \, |\xi|^2 \\ |f_{pp}(x, z, p)| \leqq M_1, \quad (x, z, p) \in \bar{\mathcal{G}} \end{gathered}$$

As before, this is seen to be no restriction.

THEOREM 9.2. Suppose f and f_p are of class C' and f satisfies (9.12) and that z is of class C' and satisfies (9.10) on G. Then z and ∇z are of class $\mathfrak{P}_2$ and z is of class C_μ' on interior domains for any μ, $0 < \mu < 1$. If, also, f and f_p are of class $C_\mu^{(n)}$ on $\mathcal{G}$, $n \geqq 1$, $0 < \mu < 1$, then z is of class $C_\mu^{(n+1)}$ on G. If, also G is of class $C_\mu^{(n+1)}$ and the boundary values of z are of class $C_\mu^{(n+1)}$, $n \geqq 1$, along G^*, then z is of class $C_\mu^{(n+1)}$ on $\bar{G}$.

PROOF. Let V be the G-potential of f_z. Since f_z is continuous and hence bounded, it follows that V is in C_μ' on $\bar{G}$, V and ∇V are in $\mathfrak{P}_2$ there and that

(9.13) $$d_2[V, C(P, r) \cap G] \leqq L_1(\mu) r^{\rho - 1 + \mu}, \quad 0 < \mu < 1$$

Also z satisfies

(9.14) $$\int_{R*} (f_{p_\alpha^i} - V^i_{x^\alpha})\, dx'_\alpha = 0$$

We then apply the device of Lichtenstein to equations (9.14) on interior domains. For all sufficiently small $h \neq 0$, we have

$$\int_{R*} (a^{\alpha\beta}_{ij} u^j_{x^\beta} + e^\alpha_i)\, dx'_\alpha = 0$$

$$l[e, C(P, r) \cap D_2] \leqq L_2(\mu) r^{\rho - 1 + \mu}, \quad 0 < \mu < 1$$

where $L_2(\mu)$ is independent of h and $a(x)$ satisfies (6.1) on $\bar{D}_2$ independently of h (the notation is that of Theorem 9.1). As usual we may let $h \longrightarrow 0$ and find that $\nabla z \in \mathfrak{P}_2$ and $C_\mu^{(o)}$ on interior domains for any μ, $0 < \mu < 1$. From this, if f and f_p are in C_μ', we first get V of class $C^2_{\mu-\epsilon}$ and a and e in $C^0_{\mu-\epsilon}$ on interior domains so $\nabla z \in C'_{\mu-\epsilon}$. Then we get V of class C^2_μ, and hence $\nabla z \in C'_\mu$. The higher regularity results are obtained by repetition.

In the case where f and f_p are of class $C_\mu^{(n)}$ on $\mathcal{G}$, and G and the boundary values of z are of class $C_\mu^{(n+1)}$ along G^*, we may first write $z = H + Z$ where $H = z$ on G^* and H is harmonic on G so $Z = 0$ on G^* and then map small boundary neighborhoods onto small hemispheres G_a as in Theorem 7.5. Both of these transformations merely replace the given f by another with the same differentiability properties.

Then the device of Lichtenstein may be applied as in the proof of Theorem 7.4 and the Euler equations used to include the higher derivatives with respect to x^{ν}.

BIBLIOGRAPHY

[1] BANACH, S., Théorie des opérations linéaires, Monografje matematyczne, Warsaw (1932).

[2] BICADZE, A. V., "On the uniqueness of the solution of the Dirichlet problem for elliptic partial differential equations," Uspekhi Matematicheskikh Nauk (N. S.) 3 (1948), 211-212.

[3] BROWDER, F. E., "The Dirichlet problem for linear elliptic equations of arbitrary even order with variable coefficients," Proceedings of the National Academy of Sciences 38 (1952), 230-235.

[4] BROWDER, F. E., "The Dirichlet and vibration problems for linear elliptic differential equations of arbitrary even order," Proceedings of the National Academy of Sciences 38 (1952), 741-747.

[5] CALKIN, J. W., "Functions of several variables and absolute continuity, I," Duke Mathematical Journal 6 (1940), 170-185.

[6] HOPF, E., "Zum analytischen Character der Lösungen regulärer zweidimensionaler Variationsprobleme," Mathematische Zeitschrift 30 (1929), 404-413.

[7] JOHN, F., "The fundamental solution of linear elliptic differential equations with analytic coefficients," Communications on Pure and Applied Mathematics 3 (1950), 273-304.

[8] JOHN, F., "General properties of the solutions of linear elliptic partial differential equations," Proceedings of the Symposium on Spectral Theory and Differential Problems, Oklahoma A. and M., Stillwater (1951), 133-175.

[9] LICHTENSTEIN, L., "Über den analytischen Character der Lösungen zweidimensionaler Variationsprobleme," Bulletin de l'Académie des Sciences de Cracovie, Classes des Sciences Mathématiques et Naturelles, sér. A (1912), 915-941.

[10] MORREY, C. B., "On the solutions of quasi-linear elliptic partial differential equations," Transactions of the American Mathematical Society 43 (1938), 126-166.

[11] MORREY, C. B., "Functions of several variables and absolute continuity, II," Duke Mathematical Journal 6 (1940), 187-215.

[12] MORREY, C. B., "Multiple integral problems in the calculus of variations and related topics," U. of California Publications in Mathematics, New Ser. Vol. 1 (1943), 1-130.

[13] SCHAUDER, J., "Über lineare elliptische Differentialgleichungen zweiter Ordnung," Mathematische Zeitschrift 38 (1943), 257-282.

[14] SCHAUDER, J., "Équations du type elliptique, problèmes linéaires," Enseignement Mathématique 35 (1935), 126-139.

[15] SCHIFFMAN, M., "Differentiability and analyticity of solutions of double integral variational problems," Annals of Mathematics 48 (1947), 274-284. (Theorem 3).

[16] VAN HOVE, L., "Sur l'extension de la condition de Legendre du calcul des variations aux intégrales multiples à plusieurs fonctions inconnues," Koninklijke Nederlandse Akademie van Wetenschappen 50 (1947), 18-23.

[17] VISIK, M. I., "On strongly elliptic systems of differential equations," Akademiia Nauk SSSR, Doklady 74 (1950), 881-884, Russian.

VIII. CONSERVATION LAWS OF CERTAIN SYSTEMS OF PARTIAL DIFFERENTIAL EQUATIONS AND ASSOCIATED MAPPINGS

C. Loewner

I would like to give a brief sketch of investiations originating in the study of a steady irrotational and compressible fluid flow in a plane.[1] If Cartesian coordinates x and y are introduced in the plane and the velocity components are called u and v, the partial differential equations governing such a flow are given by

$$(1)\qquad \begin{aligned} u_y - v_x &= 0 \\ (\rho u)_x + (\rho v)_y &= 0 \end{aligned}$$

where the density ρ is a definite function of u and v. If the differentiations indicated in (1) are performed, a system of partial differential equations of the form

$$(2)\qquad a_1^i(u,v)u_x + a_2^i(u,v)v_x + b_1^i(u,v)u_y + b_2^i(u,v)v_y = 0 \quad (i=1,2)$$

is obtained. In the original form (1) both equations of the system are conservation laws; i.e., they may be written in the form

$$(3)\qquad (\xi(u,v))_x + (\eta(u,v))_y = 0$$

with suitable continuously differentiable functions $\xi(u,v)$ and $\eta(u,v)$. The contents of such an equation may also be expressed by saying that $u = u(x,y)$, $v = v(x,y)$ are functions making

$$(4)\qquad d\chi = -\eta(u,v)dx + \xi(u,v)dy$$

an exact differential. Equations in the form of conservations laws enable us to obtain relations between the boundary values of $u(x,y)$ and $v(x,y)$

[1] For details, see the Office of Naval Research report NR-041-086 or "Conservation laws in compressible Fluid Flow and associated mappings," Journal of Rational Mechanics and Analysis, 2 (1953).

by integrating (4) over the boundary of domains in which u and v are given.

Suppose now that a system of equations of form (2) is given. We assume the coefficients a_k^i and b_k^i to be defined in an open region D of the u-v-plane and the equations, considered as linear equations for the derivatives of u and v, to be linearly independent everywhere in D. Because of the importance of conservation laws we are interested in finding all conservation laws which may be obtained by forming linear combinations of the equations of (2) with coefficients depending on u and v. It can be shown that the functions $\xi(u,v)$ and $\eta(u,v)$ describing such conservation laws are characterized as solutions of a well defined system of linear homogeneous differential equations

$$(5) \qquad c_1^i(u,v)\eta_v - c_2^i(u,v)\eta_u - d_1^i(u,v)\xi_v + d_2^i(u,v)\xi_u = 0 \qquad (i=1,2)$$

These equations may be put into the same form as the original equations by interchanging the role of dependent and independent variables:

$$(2') \qquad c_1^i(u,v)u_\xi + c_2^i(u,v)v_\xi + d_1^i(u,v)u_\eta + d_2^i(u,v)v_\eta = 0$$

It can be easily shown that a duality principle holds between systems (2) and (2') in the sense that when the operations leading from (2) to (2') are applied to (2'), we are led back to the original system (2).

We know now that system (2) can always be replaced by an equivalent system of conservation laws

$$(6) \qquad (\xi_i(u,v))_x + (\eta_i(u,v))_y = 0 \qquad (i=1,2)$$

and this can be done in many ways. To any solution of system (2) in a domain D of the x,y-plane can then be associated a pair of functions

$$(7) \qquad X_i = -\int \eta_i(u,v)dx + \xi_i(u,v)dy \qquad (i=1,2)$$

giving a mapping of D into a X_1, X_2-plane. If we take e.g., system (1) we obtain mappings into the plane of the variables representing the potential and stream functions. Important features of the system (2) are reflected in the behavior of mappings constructed in this way. We will call them mappings associated with the system (2). We wish in particular to call the attention to properties of mappings which we describe by the concepts of ellipticity or strong ellipticity of mappings and which are defined in the following way:

Let us consider a mapping of a domain D of the x,y-plane by two twice continuously differentiable functions

$$f = f(x,y)$$
(8)
$$g = g(x,y)$$

The four partial derivatives f_x, f_y, g_x, g_y will satisfy in general a well determined system of partial differential equations of the form

$$F(f_x,f_y,g_x,g_y) = 0$$
(9)
$$G(f_x,f_y,g_x,g_y) = 0$$

This system may, at given values of the arguments of F and G, be elliptic, hyperbolic, or parabolic. Setting up the relevant conditions one obtains the result:

The system (9) is elliptic, hyperbolic, or parabolic at a point (x,y) of D according to whether the quadratic differential form

$$\begin{vmatrix} df_x & df_y \\ dg_x & dg_y \end{vmatrix} = \begin{vmatrix} f_{xx}dx + f_{xy}dy, & f_{yx}dx + f_{yy}dy \\ g_{xx}dx + g_{xy}dy, & g_{yx}dx + g_{yy}dy \end{vmatrix} \tag{10}$$

is definite, indefinite, or semidefinite there.

This fact leads naturally to the

DEFINITION 1. Mapping (8) will be called elliptic, hyperbolic, or parabolic at a point (x,y) of D according to whether the quadratic differential form (10) is definite, indefinite, or semidefinite there.

Let us now consider a mapping (7) which is derived from a pair of conservation laws. The form (10) is then given by

$$\begin{vmatrix} d\xi_1(u,v) & d\eta_1(u,v) \\ d\xi_2(u,v) & d\eta_2(u,v) \end{vmatrix} \tag{10'}$$

which suggests

DEFINITION 2. A pair of conservation laws described by the function pairs $(\xi_1(u,v),$

> $\eta_i(u,v)$) $(i=1,2)$ will be called elliptic, hyperbolic, or parabolic at a point (u,v) of D according to whether the quadratic differential form (10') is definite, indefinite, or semidefinite there.

We see immediately that a solution of system (2), giving a mapping $u = u(x,y)$, $v = v(x,y)$ of D into D whose Jacobian

$$\frac{\partial(u,v)}{\partial(x,y)} \neq 0$$

in D, leads to a mapping $\chi_1 = \chi_1(x,y)$, $\chi_2 = \chi_2(x,y)$ which is elliptic, hyperbolic, or parabolic at a point (x,y) according to whether the pair of conservation laws is elliptic, hyperbolic, or parabolic at the corresponding point (u,v) of D.

Let us consider from now on only mappings (8) which are elliptic everywhere in D. We may assume that form (10) is positive definite in D. From its positivity we conclude that for a pair of points sufficiently close together the determinant of finite differences of the derivatives of f and g

$$\begin{vmatrix} \Delta f_x & \Delta f_y \\ \Delta g_x & \Delta g_y \end{vmatrix} \geqq 0 \tag{11}$$

This will, in general, not be the case for arbitrary pairs of points in D. We, therefore, introduce the useful concept of strong ellipticity of a mapping by

> DEFINITION 3. A mapping (8) will be called strongly positive elliptic in D if inequality (11) holds for arbitrary pairs of points in D. A concept of strong negative ellipticity can be defined analogously.

The advantage of introducing the concept of strong ellipticity is that assertions about the behavior of the mapping in the large may easily be made. If e.g., one of the points used in (11) is kept fixed and the other varied in D, integration of (11) over D leads to an inequality expressing the non-negativity of the area of the domain obtained from D by the mapping

$$\begin{aligned} f^* &= f - f_x(x_o,y_o)(x-x_o) - f_y(x_o,y_o)(y-y_o) \\ g^* &= g - g_x(x_o,y_o)(x-x_o) - g_y(x_o,y_o)(y-y_o) \end{aligned} \tag{12}$$

(x_0, y_0) representing the point of D which is kept fixed.

If the mapping considered is derived from a pair of conservation laws described by the function pair $(\xi_i(u,v), \eta_i(u,v))$ one is naturally led to a concept of strong ellipticity of a pair of conservation laws in form of

DEFINITION 4. The pair of conservation laws are strongly positive elliptic in a domain $\mathcal{D}$ of the u,v-plane if for any pair of points of $\mathcal{D}$ the determinant of finite differences

$$\begin{vmatrix} \Delta\xi_1(u,v) & \Delta\eta_1(u,v) \\ \Delta\xi_2(u,v) & \Delta\eta_2(u,v) \end{vmatrix} \geqq 0 \tag{13}$$

A mapping (7) derived from a solution of system (2) which assigns to any point of D a point in $\mathcal{D}$ is then necessarily strongly elliptic in D.

The indicated derivation of inequalities for strongly elliptic mappings may be utilized to obtain inequalities regarding the behavior of a subsonic compressible flow around an obstacle. One obtains e.g., (by suitable choice of the pair of conservation laws) a bound for the critical free stream Mach number in terms of geometrical data of the obstacle.[2]

[2] See "Some bounds for the critical free stream Mach number of a compressible flow around an obstacle," Anniversary Volume for Professor R. von Mises, in preparation.

IX. PARABOLIC EQUATIONS

P. D. Lax and A. N. Milgram

§1. INTRODUCTION

This paper is about the initial-boundary value problem for the parabolic equation.

(1.1) $$u_t = - Lu$$

where L is a $2p^{th}$ order elliptic differential operator

(1.2) $$Lu = (-1)^p \sum_{\nu=0}^{2p} \sum_{i_1,\ldots,i_\nu=1}^{n} a_{i_1\ldots i_\nu} \frac{\partial^\nu}{\partial x_{i_1} \cdots \partial x_{i_\nu}} u$$

The coefficients a may depend on the space variables but not on time; they are supposed to have continuous derivatives of at least p^{th} order.

The value of u at $t = 0$ is prescribed on a bounded domain G of the Euclidean space E^n; on the boundary of G the function u and all its derivatives up to order $p - 1$ are prescribed to be zero.

We shall prove in this note that this problem has a unique solution. Our proof is an application of a theorem of Hille (see [11] and [13] and Yosida (see [23]) on unbounded operators which are infinitesimal generators of semigroups (i.e., for which the exponential function e^{At} can be defined). This theorem has been applied by Hille and Yosida to the case where L is a second order elliptic operator or a system of such (see [12], [24], [25]). The application presented here for the higher order case is made possible by the recently developed theory of higher order elliptic operators (see Gårding [10] and Browder [1] and [2]) in particular by Gårding's lemma, which asserts that such operators are bounded from below for functions satisfying the first boundary condition.

The operator (1.2), as written there, applies only to $2p$ times differentiable functions (which are required to satisfy the boundary condition); it has to be extended before the Hille-Yosida theorem can be applied to it. This is accomplished here by generalizing the Friedrichs extension of symmetric half-bounded operators to the non-symmetric case. The Friedrichs extension (see [4], [5] and [19]), we recall, assigns a unique self-adjoint extension to every halfbounded symmetric operator by means of the quadratic form induced by the operator; it has been used by Friedrichs to discuss formally self-adjoint second order elliptic operators. Both Browder and Gårding make use of it in treating the first boundary value problem for formally self-adjoint higher order elliptic operators.

It is easy to verify that the extended operator $\tilde{L}$ satisfies the hypotheses of the Hille-Yosida theorem, but all we can conclude is that the <u>generalized</u> equation $u_t = -\tilde{L}u$ has a unique solution with prescribed initial value. In section 5 we show however that these generalized solutions are genuine ones, at least if the coefficients of L are sufficiently differentiable. In the case of constant coefficients we show this by use of the fundamental solution of the parabolic equation (1.1), whose properties were investigated by Ladyzhenskaya (see [17]) and P. C. Rosenbloom (see [20]). For variable coefficients the differentiability properties can be deduced from the differentiability properties of solutions of elliptic equations; such theorems were obtained recently by Gårding [10], who refers to L. Schwartz, also by Browder, who uses the fundamental singularity (see [1]), by F. John who uses the method of spherical means (see [14], [15] and [16]), and by K. O. Friedrichs who employs estimates for the L_2 norms of higher derivatives and the mollifier (see [8]). In this paper we employ Friedrichs' version of these differentiability theorems.

The extension of unsymmetric operators is presented in section 2; some further properties of this extension (which are not needed in the rest of this paper) are discussed in section 4. The Hille-Yosida theorem is described in section 3, and differentiability properties of the solution in section 5.

The initial-boundary value problem for the parabolic equation (1.1) has also been treated by F. Browder by means of eigenfunction expansions (see [3]).

Our thanks are due to the Office of Naval Research and the Office of Ordnance Research for their support.

§2. POSITIVE BILINEAR FUNCTIONALS IN HILBERT SPACE AND ASSOCIATED OPERATORS

In this section we describe the Friedrichs extension of a nonsymmetric positive definite operator. The extension is based on the theory of linear transformations induced by bilinear forms.

The following theorem is a mild generalization of the Fréchet-Riesz Theorem on the representation of bounded linear functionals in Hilbert space.

THEOREM 2.1. Let H be a real Hilbert space, and $B(x,y)$ a (not necessarily symmetric) bilinear functional which is

a) bounded, i.e.,

$$|B(x,y)| \leqq C' \|x\| \cdot \|y\|$$

and

b) positive definite in the sense that there exists a positive constant C such that

$$C \|x\|^2 \leqq B(x,x)$$

for all x in H.

If a fixed element is substituted for either argument of $B(x,y)$, B becomes a bounded linear functional of the other. We claim that all linear functionals can be obtained in this way, i.e., to each bounded linear functional l defined over H there corresponds two unique elements x_l, x_l^* in H such that

$$l(x) \equiv B(x_l,x) \equiv B(x,x_l^*)$$

for all x in H.

PROOF. Let V be the subset of H consisting of those elements y to which there corresponds an element z in H for which

$$B(z,x) \equiv (y,x)$$

for all x in H.

We note that z is unique. For suppose that z and $\bar{z}$ both satisfy this condition; then $B(z,x) = B(\bar{z},x)$ for all x in H, that is $B(z-\bar{z},x) = 0$ for all x, in particular for $x = z-\bar{z}$. Because of the positiveness of B we can conclude that $\|z-\bar{z}\| = 0$, i.e., $z = \bar{z}$.

This reasoning shows us at the same time that the dependence of z on y is <u>bounded</u> with bound $1/C$. For

$$C\|z\|^2 \leqq B(z, z) = (y, z) \leqq \|y\|\|z\|$$

Our aim is to show that all elements of H belong to V. Clearly, V is a linear subspace; furthermore from the bounded dependence of z on

y and the continuity of the bilinear functional B it follows that the linear subspace V is closed. If V were not equal to all of H there would exist an element $z^{0} \neq 0$ orthogonal to all of V. Consider the linear functional $B(z^{0},x)$. Since this is a bounded linear functional, by the Fréchet-Riesz Theorem there exists y such that $B(z^{0},x) \equiv (y,x)$. Hence y lies in V. But taking x to be z^{0} in this relation yields $B(z^{0},z^{0}) = (y,z^{0})$, and this is zero, because of the orthogonality of z_{0} to V; it follows that V is all of H.

The above reasoning shows that every ordinary scalar product can be represented as a scalar product with respect to B; since according to the Riesz-Fréchet Theorem all bounded linear functionals can be so represented this proves Theorem 2.1.

COROLLARY 2.1. If U is a proper closed linear submanifold in H, there exists x_U and x_U^* in H such that

$$B(x_U,x) \equiv B(x,x_U^*) = 0$$

for all x in U.

PROOF. Let $\bar{x}_U$ be orthogonal to U. By Theorem 2.1, there exist x_U, x_U^* such that

$$B(x_U,x) \equiv B(x,x_U^*) \equiv (\bar{x}_U,x)$$

In the applications to follow, this representation Theorem 2.1 will be applied in the case where H is a subset of another Hilbert space H_0.

We denote by (x,y), $(x,y)_0$ and $\|x\|$, $\|x\|_0$ the respective inner products and norms in H and H_0. In addition we suppose that

(2.1) H is a dense subset of H_0

(2.2) there exists a constant k such that

$$\|x\| \geqq k \, \|x\|_0$$

for all x in H.

We shall suppose as before that we are given a bilinear functional $B(x,y)$ defined on H, which, in terms of the metric in H, is bounded and positive. We shall show that this bilinear form induces two closed transformations, which are adjoints of each other:

THEOREM 2.2. There exist two linear transformations E and E^* with domains D and D^* dense in H which satisfy the relations

$$B(x,y) = (Ex,y)_o \tag{2.3}$$

for all x in D and y in H and

$$B(x,y) = (x,E^*y)_o \tag{2.4}$$

for all x in D^* and y in H.

The range of E and E^* is H_o and E and E^* have single valued inverses S and S^* which are bounded as mappings of H_o into H:

$$\|Sz\| \leqq k_1 \|z\|_o \tag{2.5}$$

they therefore are, a fortiore, bounded as mappings of H_o into H_o:

$$\|Sz\|_o \leqq k_2 \|z\|_o$$

PROOF. Each element z in H_o induces a linear functional $l(x) = (z,x)_o$ in H. By assumption (2.2) about the relation of the norms in H and H_o, it follows that $(z,x)_o$ is a bounded linear functional in H. By Theorem 2.1 there exists a unique element y in H such that $B(y,x) = (z,x)_o$ for all x in H. Denote the dependence of y on z by $y = Sz$. Clearly Sz is linear and sends all of H_o into a subset D of H. Suppose D were not dense in H. Then $\overline{D}$ is a closed linear manifold in H, and by Corollary 2.1 there exists an element x_D^* such that $B(x,x_D^*) = 0$ for all x in $\overline{D}$. But Sx_D^* lies in D, and so

$$0 = B(Sx_D^*,x_D^*) = (x_D^*,x_D^*)_o = \|x_D^*\|_o^2$$

Hence $x_D^* = 0$ and $\overline{D} = H$.

Next we show that S has a single valued inverse. For let z and v be two elements of H_o for which $Sz = Sv$; it follows by the definition of S that $(z,x)_o = (v,x)_o$ for all x in H. Thus $z - v$ is orthogonal in the sense of scalar product in H_o to a dense subset, from which we conclude $z = v$. This proves that S has a single valued inverse; call this inverse E. Clearly it satisfies (2.3).

The boundedness of S follows from the positive definite character of B, the relation of the norms in H and H_o and the Schwarz inequality:

$$C \|Sz\|^2 \leqq B(Sz,Sz) = (z,Sz)_o \leqq \|z\|_o \|E^{-1}z\|_o$$

$$\leqq \|z\|_o 1/k \|Sz\|$$

This proves the boundedness of S as mapping of H_o into H with bound $(kC)^{-1}$.

Our next theorem is concerned with possible modifications of B which do not disturb its boundedness, positiveness, or even the domain of its associated operator.

THEOREM 2.3. Let N be a linear operator with domain H and range a subset of H_o. Suppose also that

a) $|(Nx,y)_o| < M \|x\| \|y\|$ for x,y in H, and that

b) the bilinear functional $B_N(x,y) = B(x,y) + (Nx,y)_o$ is positive definite over H.

Then the operators E and E_N associated with the bilinear forms B and B_N have the same domain, and $E_N = E + N$.

PROOF. $B_N(x,y)$ is clearly bounded. Hence, B_N determines an operator E_N and domain D_N in H whose inverse S_N satisfies $B_N(S_Nz,x) \equiv (z,x)_o$ for all x in H and z in H_o. Therefore, for all x in H,

$$B(S_Nz,x) + (NS_Nz,x)_o \equiv (z,x)_o$$

or

$$B(S_Nz,x) = (z-NS_Nz,x)_o$$

By definition of S as inverse of E, this asserts that

$$S_Nz = S(z-NS_Nz) \tag{2.7}$$

This shows that S_Nz lies in D, i.e., D_N is contained in D. Denote S_Nz in (2.7) by w; apply the operator E to both sides:

$$Ew = E_Nw - Nw \tag{2.8}$$

(2.7) holds for all z in H_o, and thus (2.8) holds for all w in D_N.

Reversing the role of B and B_N we conclude that $D \subset D_N$; consequently $D = D_N$.

Let L be an unbounded operator in H_0, with domain D_0 dense in H_0. We shall call L positive definite if it satisfies these two conditions:

$$(Lx,x)_0 \geqq k(x,x)_0 \tag{2.9}$$

$$(Lx,y)_0 \leqq C'(Lx,x)_0^{1/2}\,(Ly,y)_0^{1/2} \tag{2.10}$$

The term "positive definite" is slightly misleading since it refers only to condition (2.9); "positive definite and not too unsymmetric" would be more appropriate but it is too long a phrase. We shall give a procedure which associates to each positive definite operator L an extension called its Friedrichs extension and denoted by $\tilde{L}$, and which has a bounded inverse $\tilde{L}^{-1}$ whose norm is at most k^{-1}.

Denote $(Lx,x)_0^{1/2}$ by $\|x\|$; this is a Hilbert norm over D_0 (the scalar product (x,y) is $1/2(Lx,y)_0 + 1/2(x,Ly)_0$). D_0 can be completed[1] within H_0 to a Hilbert space H under the norm $\|x\|$. $(Lx,y)_0$ is, by assumption (2.10), a bounded bilinear functional over D_0 and so can be extended by closure to a bilinear functional $B(x,y)$ over H; $B(x,y)$ satisfies the hypotheses of Theorem (2.2) (with the constant C equal to one) and so there exists a linear transformation E with domain D uniquely characterized by the requirement that $B(x,y) = (Ex,y)_0$ for all y in H and all x in D. For x in D_0, $Ex = Lx$ clearly satisfies this requirement; so E is an extension of L, and we call it[2] $\tilde{L}$. The aforementioned properties of L follow from Theorem 2.2.

---

[1] Denote by $\overline{D}_0$ the closure of D_0 under the new norm. Since every Cauchy sequence in the new norm is, by assumption 2.9, a Cauchy sequence in the old norm, there is a natural homomorphism of $\overline{D}_0$ into H_0; Friedrichs has shown, see [5], that the kernel of this homomorphism consists of the zero element, so that $\overline{D}_0$ can be regarded as a subspace H of H_0. If one operates merely with a positive definite bilinear form $B(x,y)$ over D_0, not necessarily one induced by an operator L, the kernel of the homomorphism may contain nonzero elements. Nevertheless this does not interfere with the construction of the induced operator $\tilde{L}$ (and $\tilde{L}^*$).

[2] If we operate with a bilinear form $B(x,y)$ over D_0 not induced by an operator and if the mapping T of $\overline{D}_0$ into H_0 is not one-to-one, we proceed as follows to construct $\tilde{L}$: $(z,Ty)_0$ is, for any fixed z in H_0, a bounded linear functional of y over $\overline{D}_0$. So by Theorem 2.1 it can be represented as $B(x,y)$ over $\overline{D}_0$; $x = Sz$ is a bounded transformation of H_0 into $\overline{D}_0$, so TS is a bounded mapping of H_0 into H_0. It is easy to show that this mapping has a single valued inverse, which we call $\tilde{L}$.

REMARK. $(\tilde{L}x,x)$ is equal to $\|x\|^2$.

THEOREM 2.4. Suppose that L_0 is positive definite over D_0, and that N is an operator defined over D_0. Put $L = L_0 + N$ and assume that these conditions are satisfied:

a) $(Lx,x)_0 \geqq c(L_o x,x)_o$ in D_0

b) $\|Nx\|_0 \leqq M \|x\| = M(x,L_0 x)_0$

Then $\tilde{L}$ has the same domain as $\tilde{L}_0$, and is equal to $\tilde{L}_0 + \bar{N}$, where $\bar{N}$ is the extension of N to D by closure.

PROOF. Conditions a) and b) imply that L is positive definite and that the norm induced by it is equivalent to the norm induced by L_0. So the spaces H associated with L and L_0 are the same, and L has a Friedrichs extension $\tilde{L}$. That $\tilde{L}$ is equal to $\tilde{L}_0 + \bar{N}$ follows immediately from Theorem 2.3 after we observe that condition b) enables us to extend N to H by closure.

We have now all the results that are needed for the application of the Hille-Yosida theorem. Theorem 2.4 is needed only in the special case when N is bounded by the H_0 norm, in fact when N is just a constant multiple of the identity.

§3. THE HILLE-YOSIDA THEOREM

Let L be a positive definite operator with lower bound k; if λ is a positive number, $L + \lambda I$ is positive definite with lower bound $k + \lambda$. By the fundamental property of the Friedrichs extension, $\|(\widetilde{L+\lambda I})^{-1}\|$ is less than $(k+\lambda)^{-1}$; by Theorem 2.4, taking N as λI, $(\widetilde{L+\lambda I})$ is equal to $\tilde{L} + \lambda I$. These facts can be stated in this form: The transformation $A = -\tilde{L}$ satisfies the conditions

a) The positive real axis belongs to the resolvent set of A.

b) $\|(A-\lambda I)^{-1}\|_0 \leqq (k+\lambda)^{-1}$ for positive λ.

According to the theorem of Hille-Yosida, these are just the conditions for A to be the infinitesimal generator of a strongly continuous, one-parametric semigroup of transformations which reduce to the identity of $t = 0$. More precisely:

If A satisfies conditions a) and b), there exists a one-parameter family of operators $E(t)$,

$0 \leq t < \infty$ with these properties:

(i) It forms a semigroup, $E(t_1+t_2) = E(t_1) \cdot E(t_2)$.

(ii) It is strongly continuous for $0 \leq t < \infty$

(iii) $E(0) = I$, $\|E(t)\| \leq e^{-kt}$

(iv) $\lim_{h \to 0} \frac{E(h)-I}{h} x = Ax$ for every x in the domain of A

(v) $E(t)$ commutes with $(\mu - A)^{-1}$ for all positive μ.

Note that (v) implies that $E(t)$ maps the domain of A into itself.

All this can be summarized by saying that $x(t) = E(t)x_0$ is the solution of the equation

$$\frac{dx(t)}{dt} = Ax(t)$$

with initial value x_0 (at least if x_0 lies in the domain of A). To make this paper complete we give here Yosida's elegant proof:[3]

PROOF. Define the operator A_λ as $\lambda A(\lambda - A)^{-1}$. The decomposition $A_\lambda = -\lambda + \lambda^2(\lambda - A)^{-1}$ shows that A_λ is bounded. We claim that if x belongs to the domain of A, $A_\lambda x$ tends to Ax as $\lambda \to \infty$. To see this, write $Ax = y$; $A_\lambda x$ is equal to $\lambda(\lambda - A)^{-1}y$. By condition b) we see that $\|A_\lambda x\|_0 \leq \|y\|_0$. If y belongs to the domain of A, we can write $A_\lambda x = \lambda(\lambda - A)^{-1}y = y + (\lambda - A)^{-1}Ay$; for fixed y, the second term tends to zero by condition b) and so $A_\lambda x \to y$ for y in the domain of A and hence, because of the bounded dependence of $A_\lambda x$ on y, for all y.

So far only a fraction of the information in b) was used, namely that $\|(A - \lambda I)^{-1}\|$ is $O(\lambda^{-1})$. For the next step, however, the full strength of b) is needed. We construct $E_\lambda(t) = \exp. tA_\lambda$ and we need to know that the norm of $E_\lambda(t)$ is uniformly bounded.

---

[3] See [23]. If L is symmetric, the one-parameter family of operators $E(t)$ can be constructed with the aid of the functional calculus for the bounded symmetric transformation $\tilde{L}^{-1}$; $E(t)$ is namely $\exp(-t/\tilde{L}^{-1})$. Altogether the problem of solving the initial value problem for the equation $U_t = AU$ can be regarded as part of the problem of developing a functional calculus for the operator A.

$$E_\lambda(t) = \exp.\ tA_\lambda = \exp.\ t\lambda A(\lambda - A)^{-1} =$$

$$= \exp.\ t\left\{-\lambda + \lambda^2(\lambda - A)^{-1}\right\} = \exp.(-t\lambda)\ \exp.\ t\lambda^2(\lambda - A)^{-1}$$

So $\| E_\lambda(t) \|_0 \leqq \exp.(-t\lambda)\ \exp.\ t\lambda^2 \ \|(A-\lambda)^{-1}\| \leqq \exp.(-tk\lambda(k+\lambda)^{-1})$. Next we show that $E_\lambda(t)$ converges to some operator $E(t)$ in the strong sense as λ tends to ∞. For

$$\text{(3.1)}\quad \begin{aligned} E_\lambda(t) - E_\mu(t) &= \exp.\ tA_\lambda - \exp.\ tA_\mu = \exp.[stA_\lambda + (1-s)tA_\mu]\Big|_{s=0}^{s=1} = \\ &= \int_0^1 \exp.[stA_\lambda + (1-s)tA_\mu] \cdot t[A_\lambda - A_\mu]ds \end{aligned}$$

The first factor under the integral on the right is less than one in norm (this can be shown by the same analysis that led to the inequality $\| E_\lambda \| \leqq 1$). The second factor, $[A_\lambda - A_\mu]$ applied to an element x in the domain of A tends to zero as λ and μ tend to ∞, and so $E_\lambda(t)x - E_\mu(t)x$ tends to zero for such x. Since the norm of $E_\lambda(t) - E_\mu(t)$ is uniformly bounded (at most 2), this proves the convergence of $E_\lambda(t)x$ to some limit for every x. Call this limit $E(t)x$; we claim that it has properties (i) - (v). Properties (i), (iii) follow immediately from the analogous properties of E_λ. Property (iv) follows from the integral equation

$$E(t)x = x + \int_0^t E(t)Ax\ dt$$

valid for all x in the domain of A. This integral relation is an immediate consequence of its analogue

$$E_\lambda(t)x = x + \int_0^t E_\lambda(t)A_\lambda x\ dt$$

Differentiability over the domain of A implies of course continuity over the domain of A and since $E(t)$ is uniformly bounded, this proves the strong continuity of $E(t)$.

It is not hard to show that if A is any operator satisfying the hypotheses a) and b) then there exists only <u>one</u> one-parameter family $E(t)$ satisfying conditions (i)-(v). The uniqueness proof is particularly simple in our case where $-A$ is positive definite, for this implies that for any solution $x(t)$ of

$$\frac{dx(t)}{dt} = Ax(t)$$

the norm of $x(t)$ is a decreasing function of t, and therefore two solutions of this equation with the same initial values are identical. This shows that $E(t)x$ is uniquely determined for x in the domain of A, and thus, $E(t)$ being bounded, for all x.

§4. EFFECTIVE CALCULATION OF THE FRIEDRICHS EXTENSION

In this section we shall give an effective method for computing the Friedrichs extension of a positive definite operator L which has an adjoint over its domain of definition D (and the operators occurring in most applications do have adjoints). The method relies on an extension of Theorem 2.4 to the case where, instead of requiring N to satisfy condition b), we merely require

c) $$|(Nx,y)_o| \leqq M \|x\| \cdot \|y\|$$

The positive definiteness of L is still assured and the norm induced by L is equivalent to the norm induced by L_o. So the closure of D under the norm induced either by L or L_o is the same space H; but the domains of $\tilde{L}$ and $\tilde{L}_o$ need not be the same. Nevertheless it is possible even under this weaker hypothesis to relate $\tilde{L}$ to $\tilde{L}_o$. Purely formally, we can write

$$L = L_o + N = L_o(I+L_o^{-1}N) = L_o(I+T) \tag{4.1}$$

We shall prove that (4.1) is literally true provided that L and L_o are replaced by their Friedrichs extensions $\tilde{L}$ and $\tilde{L}_o$:

$$\tilde{L} = \tilde{L}_o(I+T) \tag{4.2}$$

and if T is defined as the completion of the transformation $\tilde{L}_o^{-1}N$ over H. We start first by proving that $\tilde{L}_o^{-1}N$, defined over D_o, is bounded in the norm of H. To see this put $\tilde{L}_o^{-1}Nx = y$, which can be rewritten as $\tilde{L}_o y = Nx$ and take the scalar product of both sides with y:

$$(y,\tilde{L}_o y)_o = (y,Nx)_o$$

The left side is, by definition, equal to $\|y\|^2$; the right side is, by assumption c), less than $M\|x\| \ \|y\|$, and this proves the boundedness of $\tilde{L}_o^{-1}N$, with bound M.

Next we show that (4.2) holds for all elements x in the domain of $\tilde{L}$. Let x be such an element, with $\tilde{L}x = y$, i.e., $B(x,z) = (y,z)_o$

for all z in H. $B(x,z)$ is defined by a limiting process: $B(x,z) = \lim(Lx_i,z)_o$, where x_i is a sequence of elements in D_o tending toward x in the sense of the norm of H. The following sequence of equations uses the fact that $\tilde{L}_o$ is an extension of L_o and has an inverse defined over H_o:

$$(Lx_i,z)_o = (L_o x_i + Nx_i,z)_o = (L_o x_i,z)_o + (Nx_i,z) =$$

$$= (\tilde{L}_o x_i,z)_o + (\tilde{L}_o\tilde{L}_o^{-1}Nx_i,z)_o = (\tilde{L}_o(I+T)x_i,z)_o = B_o((I+T)x_i,z)$$

As i tends to infinity, the left term tends to $B(x,z)$ which is equal to $(y,z)_o$; the term on the right tends to $B_o((I+T)x,z)$ because of the boundedness of T and B_o. So $B_o((I+T)x,z) = (y,z)_o$ which, by definition of $\tilde{L}_o$, means that $(I+T)x$ belongs to the domain of $\tilde{L}_o$, and $\tilde{L}_o(I+T)x$ is y; this is precisely (4.2). The same argument can be used to show that every element in the domain of $\tilde{L}_o(I+T)$ belongs to the domain of $\tilde{L}$; this completes the proof of the validity of (4.2).

In a practical application it is the inverse of $\tilde{L}$ rather than $\tilde{L}$ itself that we are interested in computing, since this is the operator which relates the solution of the problem to the given data. Therein lies the usefulness of (4.2): for, if we choose L_o so that both $\tilde{L}_o^{-1}$ and $(I+T)^{-1}$ can be computed, $\tilde{L}^{-1}$ (being equal to $(I+T)^{-1}\tilde{L}_o^{-1}$) can also be computed. Such a choice of L_o is possible if L has an adjoint over D_o (i.e., an operator L^* such that $(Lx,y)_o = (x,L^*y)_o$ for x,y in D_o), namely the symmetric part of L:

$$L_o = \frac{L + L^*}{2}$$

L_o is positive definite since $(x,L_o x)_o = \frac{1}{2}(x,Lx + L^*x)_o = \frac{1}{2}(x,L^*x)_o + \frac{1}{2}(Lx,x)_o = (Lx,x)_o$ and so if L satisfies (2.9), so does L_o. (2.10) is satisfied by L_o with constant C equal to one (Schwarz inequality). Denote by N the unsymmetric part of L: $N = 1/2(L-L^*)$. Since the quadratic forms induced by L and L_o are bounded, the quadratic form induced by N is also bounded:

$$|(Nx,y)_o| \leqq M\,\|x\| \cdot \|y\| \tag{4.3}$$

The classical variational principle enables one to compute $\tilde{L}_o^{-1}$ effectively; the same is true of T, since T is the product of $\tilde{L}_o^{-1}$ and N. We shall prove now that $(I+T)^{-1}$ exists[4] and can be approximated

[4] In [1], Browder proves that if L is an elliptic operator, $I+T$ has an inverse, by showing that a) T is completely continuous (this follows from Rellich's lemma) b) $I+T$ annihilates no element of H. By the Riesz theory, a) and b) imply that $I+T$ has an inverse. Our proof does not depend on the complete continuity of T.

uniformly to any degree of accuracy by polynomials in T, and so $(I+T)^{-1}$ too can be computed effectively.

T is an antisymmetric transformation over H; this follows from the definition of scalar product in H, the symmetry of L_o and the antisymmetry of N with respect to the scalar product of H_o. Therefore $S = iT$ is a symmetric bounded transformation over H, (or rather the extension of H obtained by adjoining complex scalars). This shows right away that $I + T = I - iS$ has an inverse; this inverse can be computed with the aid of the operational calculus for bounded symmetric operators. According to this operational calculus if $f(\lambda)$ is any given continuous function over the spectrum of S, and $f_n(\lambda)$ is a sequence of polynomials tending uniformly to $f(\lambda)$ over the spectrum of S, then $f_n(S)$ tends to $f(S)$; in fact the norm of the difference of $f(S)-f_n(S)$ is not greater than the absolute value of the deviation of $f_n(\lambda)$ from $f(\lambda)$ over the spectrum of S.

As we have already shown, the norm of our operator $S = iT$ is at most M, where M is the bound of the bilinear form $(Nx,y)_o$ (in fact, it is easy to show that the norm of S -- and thus its spectral radius -- is equal to M). So to compute $(I+iS)^{-1}$ we have to approximate $f(\lambda) = (1+i\lambda)^{-1}$ uniformly by polynomials $f_n(\lambda)$ over the interval $-M < \lambda < M$. If M is less than one, this can be done by the power series expansion of $(1+i\lambda)^{-1}$ around $\lambda = 0$; for M greater than one some other method must be used.

Since the real part of $(1+i\lambda)$ is an even function of λ and its imaginary part odd, the coefficients of the even powers in $f_n(\lambda)$ will be real, those of the odd powers of λ pure imaginary; so when $f_n(S)$ is written as a polynomial in T, $f_n(S) = f_n(+iT)$, all coefficients will be real.

What we have proved so far is that $\tilde{L}^{-1} = (I+T)^{-1}\tilde{L}_o^{-1}$ can be approximated by an expression of the form $p(T)\tilde{L}_o^{-1}$, where $p(T)$ is a polynomial in T which can be picked so that $\|(I+T)^{-1}-p(T)\|$ is arbitrarily small say $< \epsilon$. (Note that since $\|(I+T)^{-1}\| = 1$, $\|p(T)\| \leq 1 + \epsilon$.) According to Theorem 2.2 and inequality (2.5), $\tilde{L}_o^{-1}$ as transformation of H_o into H is bounded, its bound being equal to k^{-1}, the reciprocal of the lower bound (2.9) of the operator L_o. So if x is any element of H_o, $p(T)\tilde{L}_o^{-1}x$ differs from $\tilde{L}^{-1}x$ in the sense of the norm of H by at most $\epsilon k^{-1}\|x\|_o$.

For sake of completeness we state the variational principle for positive definite symmetric operators:

Let x be any element of H_o, and let y vary in the domain of $\tilde{L}_o$. Consider the function $V(y) = (\tilde{L}_o y,y)_o - 2(x,y)_o$; it reaches its minimum for $y = \tilde{L}_o^{-1}x$. Denote the value of this minimum by d. If y' is any other element in the domain of $\tilde{L}_o$, the deviation of $V(y')$ from

d measures the distance of y' from y in the sense of the norm of H:

$$\| y'-y \|^2 = V(y') - d$$

This shows that a sequence of elements y_i approaches y if and only if $V(y_i)$ approaches d. This is the basis of the Rayleigh-Ritz procedure which does effectively produce a sequence of elements of D_0 tending to y in the sense of the norm of H (note that in order to estimate the distance of y_i from y, we need a <u>lower</u> bound for d).

Suppose x is any given element of H_0 and we wish to evaluate $\tilde{L}^{-1}x$ to some desired degree of accuracy. According to what was said before, $\tilde{L}^{-1}x$ differs from $p(T)\tilde{L}_0^{-1}x$ by less than $\epsilon k^{-1} \| x\|_0$. To evaluate $p(T)\tilde{L}_0^{-1}x$, we find by the Rayleigh-Ritz method an approximation to $\tilde{L}_0^{-1}x$, i.e., an element y' of D_0 whose distance from $\tilde{L}_0^{-1}x$ in the sense of the H norm is less than η. $p(T)\tilde{L}_0^{-1}x$ differs from $p(T)y'$ by less than $\|p(T)\| \cdot \eta \leqq (1 + \epsilon)\eta$ To evaluate $p(T)y'$ to any desired degree of accuracy, we have to be able to evaluate terms of the form $T^n y'$ to any desired degree of accuracy; we shall show how to evaluate Ty': Since y' lies in D_0, the original domain of definition of $T = \tilde{L}_0^{-1}N$, we simply form $Ny' = z$ and approximate $\tilde{L}_0^{-1}z$ by the Rayleigh-Ritz method to an arbitrary degree of accuracy by an element of D_0.

We close this section by stating a result on positive definite operators with an adjoint; this result will be used in Section 5:

> If L is a positive definite operator defined over D_0 with an adjoint L^* over D_0, then $\tilde{L}$ and $\tilde{L}^*$ are the adjoints of each other.

This result is a corollary of Theorem 2.2. Actually, the only part of it needed in the next section is that $\tilde{L}$ and $\tilde{L}^*$ are adjoint to each other, i.e., $(\tilde{L}x,y) = (x,\tilde{L}^*y)$ for all x in D and all y in D_0. This property of $\tilde{L}$ is sometimes expressed by saying that $\tilde{L}$ is a <u>weak</u> extension of L.

§5. DIFFERENTIABILITY PROPERTIES OF SOLUTIONS

The $2p^{th}$ order operator L of (1.2) is called <u>elliptic</u> if the $2p^{th}$ order form

$$\sum_{i_1,\ldots,i_{2p}=1}^{n} a_{i_1\ldots i_{2p}}\, \xi_{i_1}\ldots\xi_{i_{2p}} \tag{5.1}$$

does not vanish (say it is positive) for any choice of the variables ξ except $\xi_1 = \ldots \xi_n = 0$. For the domain D_o of this operator we take all 2p times differentiable functions which vanish outside of a given bounded domain G. Denote by $(u,v)_o$ and $\| u \|_o$ the L_2 scalar product and norm of the function u over G and by $\| u \|_k$ the sum of the L_2 norms of u and its first k derivatives. We define H_o as the closure of the space of continuous square integrable functions u over G under the norm $\| u \|_o$; D_o is a dense subspace of H_o. For elliptic operators whose highest order coefficients are constant, L. Gårding has proved by the use of the Fourier transform (see [9]) that if L is augmented by a suitably large multiple of the identity, the resulting operator $L + t_o I$ is positive definite; specifically that these inequalities hold:

$$(5.2)\qquad \begin{aligned} (Lu+t_o u,u)_o &\geqq h \ \| u \|_p^2 \\ (Lu+t_o u,v)_o &\leqq C' \ \| u \|_p \ \| v \|_p \end{aligned}$$

In [10], Gårding has announced the same result for elliptic operators with variable coefficients; it can be deduced from the result for the constant coefficient case with the aid of a well-known estimate for $\| u \|_k$, $k < p$, in terms of $\| u \|_o$ and $\| u \|_p$, and a sufficiently fine partition of unity. (See for instance F. Browder, [2] or J. Leray, [18]; Leray treats the more general case of strongly elliptic systems.)

Since, for the purpose of solving the parabolic equation

$$\frac{\partial u}{\partial t} = - Lu$$

the addition of a constant multiple of the identity to L is irrelevant, we may as well assume that already L itself satisfies (5.2):

$$(5.2')\qquad \begin{aligned} (Lu,u)_o &\geqq h \ \| u \|_p^2 \\ (Lu,v)_o &\leqq C' \ \| u \|_p \ \| v \|_p \end{aligned}$$

which implies that L satisfies (2.9) and (2.10), i.e., L is a positive definite operator.

According to sections 2 and 3, our elliptic operator L has a Friedrichs extension $\tilde{L}$, and $-\tilde{L} = A$ is the infinitesimal generator of a semigroup $E(t)$, i.e., $E(t)u_o = u(t)$ satisfies $du(t)/dt = -\tilde{L}u(t)$, $u(0) = u_o$ for every u_o in the domain of A.

$u(t)$ is a generalized solution of our parabolic equation with initial value u_o. We shall prove now that $u(t)$ is a genuine solution; i.e., that $u(t)$ has 2p continuous derivatives with respect to the space variables and continuous derivatives of all orders with respect to t at every point of the domain D and all time $t > 0$, and that it satisfies the partial differential equation $u_t = - Lu$ there, provided that the coefficients of the operator L are sufficiently differentiable. Further differentiability of the coefficients implies further differentiability of $u(t)$ with respect to the space variables. We shall prove these statements for any initial function in H_o and not just for elements of the domain of A; furthermore we shall show that if the initial function u_o is sufficiently differentiable, $u(t)$ approaches u_o as t tends to zero not only in the L_2 sense but pointwise.

It should be pointed out that from the relation $u_t = - \tilde{L}u$ and differentiability properties of $u(t)$ we can deduce that the partial differential equation $u_t = - Lu$ is satisfied, on the basis of this simple result:

> If u is a function with continuous derivatives up to order $2p$ which belongs to the domain of $\tilde{L}$, then $\tilde{L}u$ is indeed equal to Lu, L being interpreted as the originally given differential operator.

This is not an immediate consequence of the definition of $\tilde{L}$, since $\tilde{L}$ is the Friedrichs extension of the operators L as defined over D_o, D_o being the set of $2p$ times differentiable functions which satisfy the boundary conditions. Nevertheless the result is true as stated; its proof is based on the fact, noted at the end of section 4, that $\tilde{L}$ is a weak extension of L. Denote namely $\tilde{L}u$ by v, Lu by v', and take w as some function with $2p$ continuous derivatives in D which vanishes outside some closed subset of D. w belongs to the domain of L^* (in order that L^* exist the coefficients of the terms of order ν, $\nu = 0,1,\ldots,2p$ have to be ν-p times differentiable) and since $\tilde{L}$ is a weak extension, $(v,w) = (\tilde{L}u,w) = (u,L^*w)$. On the other hand, by Green's formula, $(v',w) = (Lu,w)$ is also equal to (u,L^*w); this shows that v' is square integrable over D and equal to v.

Our proof of the differentiability properties of $u(t)$ is based on a differentiability theorem of K. O. Friedrichs (see [8]) for solution of elliptic equations. The reduction is via this lemma:

> LEMMA 5.1. For t positive, $E(t)$ has derivatives of arbitrary order with respect to t, and $d^mE(t)/dt^m$ is equal to $A^mE(t)$, $m = 1,2,\ldots,$. The differentiation with respect to t is meant in the uniform topology.

A consequence of Lemma 5.1 is that $E(t)$ transforms every element of H_0 into an element to which the operator A can be applied an arbitrary number of times; since it follows from the above mentioned theorem of Friedrichs that such elements are as differentiable as the coefficients of L permit them to be, this proves the differentiability of $E(t)u_0$ with respect to the space variables. To be precise:

> FRIEDRICHS THEOREM. Let u be an element in the domain of A and denote Au by v. If v has square integrable derivatives up to order S in every closed subset of G, then u has square integrable derivatives up to order $2p + S$ in every closed subset of G, and so -- by Sobolev's lemma (see [22])-- continuous derivatives of order $2p + S - [n/2] - 1$ in G. The coefficients of L have to have continuous derivatives up to order $p + S$.

> COROLLARY. If u lies in the domain of A^m, u has continuous derivatives up to the order $2mp - [n/2] - 1$ in G, provided that the coefficients of L have continuous derivatives up to the order $(2m-1)p$.

Combining this corollary with Lemma 5.1 we conclude that if the coefficients of L have $p + [n/2] + 1$ continuous derivatives, all elements of the form $E(t)u$ have continuous derivatives up to order $2p$ with respect to the space derivatives, and if the coefficients have derivatives of arbitrary order, so do all elements of the form $E(t)u$.

To prove the pointwise differentiability of $E(t)u$ with respect to t, we need the full statement of Friedrichs theorem; we shall state it as a corollary to the corollary:

> COROLLARY^2. A^{-m} not only transforms every element u of H_0 into a function with $2mp - [n/2] - 1$ continuous derivatives but the magnitude of these derivatives is bounded by the L_2 norm of u, uniformly in every closed subset of G.

Choose m_0 so large that $2m_0p - [n/2] - 1$ is non-negative; by the second corollary A^{-m_0} maps every sequence of elements of H_0 convergent in the norm of H_0 into a sequence of functions uniformly convergent in every closed subset of G.

Let u_0 be any element of H_0, t_1 any positive value: Choose t_2 as some smaller positive number; using the semigroup property of $E(t)$, the fact that $E(t_2)u_0$ lies in the domain of A^{m_0} (Lemma 5.1) and that A^{-m_0} and $E(t)$ commute (property (v)) we can write:

$$E(t_1)u_0 = E(t_1-t_2)\,E(t_2)u_0 = E(t_1-t_2)A^{-m_0}A^{m_0}\,E(t_2)u_0 =$$

$$= A^{-m_0}\,E(t_1-t_2)v_0$$

where v_0 is an abbreviation for $A^{m_0}E(t_2)u_0$. The same process applied to $t_1 + \delta$ in place of t_1 yields a similar identity; subtracting the two and dividing the difference by δ we obtain

$$(5.3)\qquad \frac{E(t_1+\delta)u_0 - E(t_1)u_0}{\delta} = A^{-m_0}\,\frac{E(t_1+\delta-t_2)v_0 - E(t_1-t_2)v_0}{\delta}$$

$E(\tau)v_0$ is differentiable at $\tau = t_1 - t_2$, and so the function to the right of A^{-m_0} on the right side of (5.3) tends to a limit in the sense of H_0, as δ tends to zero. According to what was said before, the images of these functions under the transformation A^{-m_0} converge uniformly in every closed subset of G; this proves the pointwise differentiability of $E(t)u_0$ at $t = t_1$.

By an entirely analogous reasoning one can show that all higher t derivatives of $E(t)u_0$ exist pointwise; no further differentiability conditions need be imposed on the coefficients of L. This type of argument also shows that if u_0 is sufficiently smooth, i.e., belongs to the domain of A^m, $E(t)u_0$ tends to u_0 as t tends to zero not only in the sense of the norm of H_0 but pointwise as well.

Now we turn to the proof of Lemma 5.1; it rests on a contour integral representation for $E(t)$. This powerful method for studying semigroups has been developed by E. Hille (see [11]).

We start with the remark that if A is an operator that satisfies the hypotheses of the Hille-Yosida theorem, then not only the positive real axis but the whole right half plane $\mathrm{Re}\{\mu\} > 0$ belongs to the resolvent set of A and the norm of the resolvent, $\|(A-\mu I)^{-1}\|_0$, is bounded by $[\mathrm{Re}\{\mu\}]^{-1}$ there.[5] On the strength of this an integral representation can be derived for $E(t)$ (or at least for $A^{-1}E(t)$) but not one from which the differentiability of $E(t)$ with respect to t could be deduced. To deduce that we need to know more about the resolvent

[5] This can be deduced from the resolvent equation $A - \mu I = (A-\lambda I)\left\{I + (\lambda+\mu)(A-\lambda I)^{-1}\right\}$, λ large positive.

set[6] of the operator A:

LEMMA 5.2. If L is a $2p^{th}$ order elliptic operator, $\tilde{L}$ its Friedrichs extension, and $A = -\tilde{L}$, then all complex numbers $\mu = \alpha + i\beta$ in the left half plane for which

$$|\beta| \geq \text{const}\, |\alpha|^{\frac{2p-1}{2p}}$$

holds belongs to the resolvent set, and the norm of the resolvent at such a point is bounded by

$$\|(A-\mu I)^{-1}\|_o \leq \text{const}\, |\beta|^{-1}$$

A proof of this lemma will be given in a forthcoming paper by J. Berkowitz and P. D. Lax; it is based on the Gårding inequality, a well-known inequality[7] for $\|u\|_k$, $k < p$, in terms of $\|u\|_p$ and $\|u\|_o$ and the Friedrichs extension (in a slightly more general setting than was needed in section 2).

The contour integral representation for $E(t)$ is

$$(5.4) \qquad E(t) = (2\pi i)^{-1} \int (\mu - A)^{-1} e^{\mu t}\, d\mu$$

the contour being the curve

$$|\beta| = \text{const}\, |\alpha|^{\frac{2p-1}{p}} \qquad -\infty < \alpha \leq 0$$

(or any other contours in the left half plane). For positive t the factor $e^{\mu t}$ tends to zero exponentially as $\mu \longrightarrow \infty$, while the first factor $(\mu - A)^{-1}$ is

$$O\left(|\mu|^{-\frac{2p-1}{p}}\right)$$

---

[6] Of course if L is self-adjoint all complex numbers with nonzero imaginary parts belong to the resolvent set, and $\|(A-\mu I)^{-1}\| \leq |\beta|^{-1}$. But in this case no integral representation is needed to deduce the differentiability of $E(t)$ with respect to t; this follows from the functional calculus for the bounded symmetric transformation $\tilde{L}^{-1}$.

[7] $\|u\|_k^2 \leq C\, a^{p/(p-k)}\, \|u\|_o^2 + \frac{1}{a}\, \|u\|_p^2$, where C is a constant depending on the indices p, k and the domain; and a is any positive number.

so the integral on the right of (5.4) converges in the uniform topology. If we differentiate the right side of (5.4) with respect to t under the integral sign any number of times, the resulting integral still converges in the uniform topology; this shows that the right side of (5.4) has derivatives of arbitrary order with respect to t.

Denote the object on the right side of (5.4) by $F(t)$ temporarily. We shall prove that $F(t) \equiv E(t)$ by showing that

a) $$\frac{dF(t)}{dt} = AF(t)$$

b) $F(t)u_0 \longrightarrow u_0$ as $t \longrightarrow 0$ for all u_0 in the domain of A.

According to the remarks at the end of section 3, a) and b) imply that $F(t)u_0$ is equal to $E(t)u_0$ for all t and all u_0 in the domain of A and so, by continuity, for all u_0 in H_0.

To prove a), we note that the operator A can be applied to the integrand in (5.4) and that the resulting integral converges absolutely. Since A is a closed operator this implies that

$$AF(t) = (2\pi i)^{-1} \int A(\mu - A)^{-1} e^{\mu t} \, d\mu$$

On the other hand, dF/dt is given by

$$\frac{d\,F(t)}{dt} = (2\pi i)^{-1} \int u(\mu - A)^{-1} e^{\mu t} \, d\mu$$

The difference of these two is $(2\pi i)^{-1} \int e^{\mu t} \, d\mu$, and the value of this integral is zero, as may be seen by shifting the contour of integration to the left. This shows that dF/dt is equal to AF; the same type of reasoning shows that d^mF/dt^m is equal to A^mF.

We turn now to the proof of b). Let u_0 be any element in the domain of A; choose for λ_0 any complex number to the <u>right</u> of the contour of integration in (5.4), say $\lambda_0 = 1$; denote $(A - \lambda_0)u_0$ by v_0.

$$F(t)u_0 = F(t)(A - \lambda_0)^{-1}v_0 =$$

$$(2\pi i)^{-1} \int (A - \lambda_0)^{-1} A - (\mu - A)^{-1} v_0 e^{\mu t} \, d\mu$$

We claim that the right side is equal to

$$(2\pi i)^{-1} \int (\mu - \lambda_0)^{-1} (A - \mu)^{-1} v_0 \, e^{\mu t} \, d\mu$$

This is most easily verified by showing that the difference between the two expressions which we claim to be equal,

$$(2\pi i)^{-1} \int (A-\lambda_0)^{-1} (\mu-\lambda_0)^{-1} e^{\mu t} \, d\mu$$

is zero. Since the only singularity of this integral is at $\mu = \lambda_0$, we may shift the contour to the left without altering the value of the integral; but we see that, on the new contour, the integral is arbitrarily small.

The representation

$$(2\pi i)^{-1} \int (\mu-\lambda_0)^{-1} (A-\mu)^{-1} v_0 \, e^{\mu t} \, d\mu = F(t)u_0$$

was derived in order to determine the limit of $F(t)u_0$ as t tends to zero. For that purpose note that the integral converges absolutely and uniformly for all t, including $t = 0$, since the first factor is $O(|\mu|^{-1})$ near infinity, the second factor - according to Lemma 5.2 -

$$O\left(|\mu|^{-\frac{2p-1}{2p}}\right)$$

and the third, $e^{\mu t}$, never exceeds one in absolute value. This justifies passage to the limit under the integral sign, and so

$$\lim_{t\to 0} F(t)u_0 = (2\pi i)^{-1} \int (\mu-\lambda_0)^{-1}(A-\mu)^{-1} v_0 \, d\mu$$

To evaluate this remaining integral we make a closed path out of the contour by adjoining the arc of the circle $|\mu| = R$ which is to the right of the contour, and throwing away that portion of the old contour which lies outside of this circle. The value of the integral along the new arc and the discarded arc is <u>small</u> if R is large; therefore this change makes no difference in the value of the integral, which can now be put equal to the sum of the residues inside. The only singularity inside is at $\mu = \lambda_0$, and the residue there is $(A-\lambda)^{-1} v_0 = u_0$. Q.E.D.

We should like to mention that it is possible to prove differentiability theorems for generalized solutions of a parabolic equation directly, without the aid of differentiability theorems for solutions of elliptic equations. We have carried out such a direct verification in case the coefficients of L are constant, using the fundamental solution $K(x,t)$ for our parabolic equation.[8] The fundamental solution in this case can be written down explicitly; denoting by $L^*(\xi)$ the form

---

[8] Yosida, in [26], uses such a method to prove the differentiability of generalized solutions of the diffusion equation, even in the case of variable coefficients.

$$L^*(\xi) = \sum_{\nu=0}^{2p} \sum_{i_1,\dots,i_\nu=1}^{n} (-1)^{p-\mu} a_{i_1,\dots,i_\nu} \xi_{i_1},\dots,\xi_{i_\nu}$$

we have

$$K(x,t) = \int_{E_n} \exp\left\{2\pi i\xi \cdot x - tL^*(2\pi i\xi)\right\} d\xi$$

The integral defining $K(x,t)$ converges absolutely and uniformly for $T \geqq t_o > 0$. Various properties of $K(x,t)$ were investigated by Ladyzhenskaya, [17] and P. C. Rosenbloom, [20]. The ones we need to know are these:

a) $K(x,t)$ satisfies $\frac{\partial K}{\partial t} = - L^*K$ for $t > 0$.

b) $K(x,t)$ has derivatives of arbitrary order with respect to x and t for $t > 0$; as t tends to zero, K and its derivatives remain bounded (and in fact tend to zero) uniformly outside of any sphere $\| x \| = r$.

c) $u(y,t) = (K(x-y,t),u)_o$ tends to $u(y)$ in the sense of the norm of H_o as $t \longrightarrow o$.

Let x_o be any point of G, $\psi(x)$ some function identically equal to one in some neighborhood N of x_o, zero outside of some closed subset of G and 2p times differentiable. Let u_o be some element in the domain of $\tilde{L}$, so that $E(\tau)u_o = u(\tau)$ satisfies $u_\tau = - \tilde{L}u$. Take the scalar product of both sides with $\psi(x)K(x-y,t-\tau)$:

$$(u_\tau, \psi K) = (-\tilde{L}u, \psi K)$$

Since $\tilde{L}$ is a weak extension, and ψK lies in the domain of L^*, the right side can be written as $(u,-L^*\psi K)$; the left side can be written as $(u,\psi K)_\tau - (u,\psi K_\tau)$ which is equal to $(u,\psi K)_\tau + (u,\psi K_t)$, since K is a function of $t - \tau$. Integrating both sides with respect to τ from zero to $t - \delta$, we get:

$$\text{(5.5)}\quad \Big(u(t-\delta),\, \psi(x)K(x-y,\delta)\Big)_o = \Big(u,\, \psi(x)K(x-y,t)\Big)_o - \int_o^{t-\delta} (u, \psi K_t + L^*\psi K)_o \, d\tau$$

The factor ψ is $\equiv 1$ in N; so by property a) of K, $\psi K_t + L^*\psi K = K_t + L^*K$ is zero in N. This shows that the space integration in the second term of (5.5) is extended only over G - N; so, by property b) of K, the function on the right side of (5.5) is an infinitely differentiable function of y and t and its derivatives remain uniformly bounded as δ tends to zero (i.e., uniformly with respect to δ, not the order of the derivative) if y is restricted to some closed subset of N. So the limit of the right side as $\delta \longrightarrow 0$ as function of y, y in N, is an infinitely differentiable function. The limit of the left side, regarded as element of H_0, is by property c) of K equal to $\psi u(t)$. This shows that u(t) is an infinitely differentiable function in N, i.e., (N being arbitrary) in all of G.

The methods of section 5 can be used to derive analogous results in case u is a vector of functions and L is a strongly elliptic matrix differential operator.

BIBLIOGRAPHY

[1] BROWDER, F. E., "The Dirichlet problem for linear elliptic equations of arbitrary even order with variable coefficients," Proceedings of the National Academy of Sciences, 38 (1952), 230-235.

[2] BROWDER, F. E., "The Dirichlet and vibration problems for linear elliptic differential equations of arbitrary order," Proceedings of the National Academy of Sciences, 38 (1952), 741-747.

[3] BROWDER, F. E., "Linear parabolic equations of arbitrary order," abstract 605, Bulletin of the American Mathematical Society, 58 (1952), 632.

[4] FREUDENTHAL, H., "Ueber die Friedrichssche Fortsetzung halbbeschränkter Hermitescher Operatoren," Proceedings of the Koninklijke Nederlandsche Akademie van Wetenschappen, 39 (1936), 832-833.

[5] FRIEDRICHS, K. O., "Spectraltheorie halbbeschränkter Operatoren und Anwendung auf die Spektralzerlegung von Differzialoperatoren I, II," Mathematische Annalen, 109 (1934), 465-486, 685-713.

[6] FRIEDRICHS, K. O., "On differential operators in Hilbert space," American Journal of Mathematics 61 (1939), 523-544.

[7] FRIEDRICHS, K. O., "The identity of weak and strong extensions of differential operators," Transactions of the American Mathematical Society 55 (1944), 132-151.

[8] FRIEDRICHS, K. O., "Differentiability of solutions of elliptic operators," Communications on Pure and Applied Mathematics, 6, 3.

[9] GARDING, L., "Dirichlet's problem and the vibration problem for linear elliptic partial differential equations with constant coefficients," Proceedings of the symposium on spectral theory and differential problems, Stillwater, Oklahoma, (1951), 291-301.

[10] GARDING, L., "Le problème de Dirichlet pour les équations aux dérivées partielles elliptiques linéaires dans des domaines bornes," Comptes rendus des séances de l'Académie des Sciences, 233 (1951), 1554-1556.

[11] HILLE, E., Functional Analysis and Semi-Groups, American Mathematical Society Colloquium Publication, New York, (1948).

[12] HILLE, E., "On the integration problem for Fokker-Planck's equation," 10th Congress of Scandinavian Mathematicians, Trondheim (1950).

[13] HILLE, E., "On the differentiability of semi-group operators," Acta Scientiarum Mathematicarum, 12, part B (1950), 19-26.

[14] JOHN, F., "Derivatives of continuous weak solutions of linear elliptic equations," Communication on Pure and Applied Mathematics, 6, 3.

[15] JOHN, F., "General properties of solutions of linear elliptic partial differential equations," Proceedings of the symposium on spectral theory and differential problems, Stillwater, Oklahoma (1951).

[16] JOHN, F., "Derivatives of solutions of linear partial differential equations," this Study.

[17] LADYZENSKAYA, O. A., "On the uniqueness of the solution of the Cauchy problem for a linear parabolic equation," Matematiceskii Sbornik, 27, 69 (1950), 175-184.

[18] LERAY, J., "Hyperbolic equations with variable coefficients," lectures delivered at the Institute for Advanced Study, Princeton, Fall, (1952).

[19] NAGY, B. v. Sz., Spectraldarstellung linearer Transformationen des Hilbertschen Raumes, Springer, Berlin (1942).

[20] ROSENBLOOM, P. C., Linear equations of parabolic type with constant coefficients," this Study.

[21] ROTHE, E., "Ueber die Wärmeleitungsgleichung mit nichtkonstanten Koeffizienten im räumlichen Falle I, II," Mathematische Annalen 104 (1931), 340-354, 354-362.

[22] SOBOLEV, S., "Sur quelques évaluations concernant les familles des fonctions ayant des dérivées à carré intégrables, Akademiya Nauk USSR, Doklady, N.S. (Comptes Rendus) (1938), 279-282.

[23] YOSIDA, K., "On the differentiability and the representation of one-parameter semigroups of linear operators," Journal of the Mathematical Society of Japan, 1 (1948), 15-21.

[24] YOSIDA, K., "Integration of Fokker-Planck's equation in a compact Riemann space," Arkiv for Matematik, 1 (1949), 71-75.

[25] YOSIDA, K., "Integration of the Fokker-Planck equation with boundary conditions," Journal of the Mathematical Society of Japan, 3 (1951).

[26] YOSIDA, K., "On the integration of diffusion equations in Riemannian spaces," Proceedings of the American Mathematical Society, 3 (1952), 864-874.

[27] VISCHIK, M. I., "The method of orthogonal and direct decomposition in the theory of elliptic differential equations," (Russian), Matemakiceskii Sbornik, 25 (1949), 189-234.

[28] WEYL, H., "The method of orthogonal projection in potential theory," Duke Mathematical Journal, 7 (1940), 414-444.

X. LINEAR EQUATIONS OF PARABOLIC TYPE WITH CONSTANT COEFFICIENTS

P. C. Rosenbloom

In this paper we are concerned with equations of the form

$$\frac{\partial u}{\partial t} = L\left(\frac{\partial}{\partial x}\right)u \tag{1}$$

Here $u = u(x_1, \ldots, x_n)$ is a function on the Euclidean space E_n of n dimensions,

$$\frac{\partial}{\partial x}$$

is the symbolic vector

$$\left[\frac{\partial}{\partial x_1}, \ldots, \frac{\partial}{\partial x_n}\right]$$

and $L(\xi)$ is a polynomial in the vector $\xi \in E_n$ with constant coefficients, which may be arbitrary complex numbers. Let m be the degree of L, let q be the homogeneous part of degree m, and let $L = q + r$, where $\deg r < m$. We shall say that L is <u>elliptic</u> if $\operatorname{Re}\{q(\xi)\}$ is definite for real ξ. In that case $m = 2k$ must be even; we say that (1) is <u>parabolic</u> if $\operatorname{Re}\left\{(-1)^{k-1}q(\xi)\right\}$ is positive definite for real ξ.

By standard Fourier transform techniques one finds that the fundamental solution of (1) is

$$K(x,t) = \frac{1}{(2\pi)^n} \int_{E_n} \exp\{i(\xi \cdot x) + tL(i\xi)\}\, dV_\xi \tag{2}$$

where dV_ξ is the volume element of E_n, ξ denoting the variable of integration. In terms of (2) we solve formally the initial value problem:

$$\begin{aligned} &u \text{ satisfies (1) for } t > 0, \text{ all } x \\ &u(x,0) = \phi(x) \end{aligned} \tag{3}$$

by means of the Poisson integral formula

(4) $$u(x,t) = \int \phi(y)\ K(x-y,t)dV_y = (T_t\phi)\ (x)$$

Our present purpose is to discuss in what sense and to what extent (4) is the solution of (3), and as a byproduct to study the properties of functions represented by (4).

The treatment of (3) by Fourier transforms (see Titchmarsh [14], Bochner and Chandrasekharan [2]) leads to (4) directly, but yields only very crude conditions for the validity of (4) and for the uniqueness of the solution of (3). For the treatment of more delicate problems, a preliminary investigation of the function (2) is needed as well as some elementary properties of functions of the form (4).

If (1) is parabolic, then clearly the integral in (2) is absolutely convergent for $\operatorname{Re}\{t\} > 0$ and analytic for all x and t such that $\operatorname{Re}\{t\} > 0$. The following are direct consequences of the representation of K as a Fourier transform:

$$\int_{E_n} K(x,t)\ dV_x = e^{L(0)t} = e^{a_o t}, \qquad \operatorname{Re}\{t\} > 0$$

$$\int_{E_n} K(x-y,t_1)\ K(y,t_2)dV_y = K(x,t_1+t_2), \qquad \operatorname{Re}\{t_j\} > 0$$

The second equation expresses the semigroup property of the transformations $\{T_t\}$ and is one form of Huygen's principle.

The following estimates are due to Ladyzhenskaya [5]:

(5)
$$|K(x,t)| \leqq C_1 t^{-n/m} \exp\ (C_2 t - C_3\ \|x\|^{\mu}\ t^{-(m-1)})$$

$$|D_x^v K| \leqq C_1(v)\ t^{-(n+v)/m} \exp\ (C_2 t - C_3\ \|x\|^{\mu}\ t^{-(m-1)})$$

where

$$\|x\|^2 = \sum_{k=1}^{n} |x_k|^2, \qquad \mu = m/(m-1)$$

and D_x^v is an arbitrary derivative of order v with respect to the variables $x_1,\dots,x_n$. The constants which appear in (5) can be given the following explicit expressions (see [12]):

$$
(6)\qquad
\begin{aligned}
C_1 &= I_0(2\pi)^{-n}\ (2/h)^{n/m} \\
C_2 &= B\left\{2^{m-1} + 1 + m^{-1}\ 2^{m^2+m-2}\ \ (B/h)^{m-1}\right\} \\
C_3 &= (m-2)\ m^{-\mu}\ C_0^{\ -1/(m-1)} \\
C_1(v) &= 2^v(2\pi)^{-n}(I_v+I_0(v/e)^{v/n}\ (h/2m\ C_0^{\ v/m})\ (2/h)^{(v+n)/m}
\end{aligned}
$$

where

$$
\begin{aligned}
A &= \max\ |q(\xi+i\eta) - q(\xi)|, && \|\xi\| \leqq 1,\ \ \|\eta\| \leq 1, && \xi,\ \eta \text{ complex} \\
B &= \max\ |r(\xi)| && \|\xi\| \leqq 1, && \xi \text{ complex} \\
h &= \min \operatorname{Re}\ \{(-1)^{k-1}\ q(\xi)\} && \|\xi\| = 1, && \xi \text{ real}
\end{aligned}
$$

$$
C_0 = A\left\{1 + m^{-1}\ 2^{m-2}\ (A/h)^{m-1}\right\} + 2^{m-1}\ B
$$

$$
I_v = 2\pi^{n/2}\ \Gamma((v+m)/m)/(m\,\Gamma(n/2))
$$

Such estimates are useful in the study of equations with variable coefficients.

For the case $n = 1$, $L = q$, estimates like (5) were obtained by Täcklind [13]. Polya, in his diary, has obtained an asymptotic development of K in this case by the saddle point method. Turrittin has shown the author how such an asymptotic series can also be obtained by the methods of Birkhoff and Trjitzinsky [15]. Probably the analogous results for the general case can be obtained by the methods of Van der Corput.

We now give a few results concerning functions represented by (4).

THEOREM 1. If

$$
\int |\phi(z)|\ \exp\ (-M\ \|z\|^{\mu}\ )dV_z = N < +\infty
$$

then $u = T_t\phi$ exists and satisfies (1) for $0 < t < (C_3/M)^{m-1}$. If $0 < \beta < 1$,

$$
\alpha = (1 - \beta^{m-1})^{-(\mu-1)}
$$

then

$$
\int |u(x,t)|\ \exp\left\{-\alpha M\ \ \|x\|^{\mu}\right\}\ dV_x \leqq C_1\ N\ C_3^{\ -n/\mu}\ I_0(1-\beta)^{-n/\mu}\ e^{C_2 t}
$$

for

$$
0 < t \leqq (\beta^2 C_3/M)^{m-1}
$$

Also

$$|u(x,t)| \leq NC_1 t^{-n/m} \exp\left\{ C_2 t + \frac{M\|x\|^{\mu}}{[1-t(M/C_3)^{m-1}]^{1/(m-1)}} \right\}$$

for $0 < t < (C_3/M)^{m-1}$. We have

$$\lim_{t \to 0+} u(x,t) = \phi(x)$$

at all points of the Lebesgue set of ϕ, and

$$\lim_{(x,t) \to (x_0, 0+)} u(x,t) = \phi(x_0)$$

at all points of continuity of ϕ.

For the case of the heat equation

$$L(\xi) = \sum_{k=1}^{n} \xi_k^2$$

these results are also true, with obvious modifications, for functions representable as Poisson-Stieltjes integrals. This is probably true in general. It is not difficult to estimate the rate of convergence of u to ϕ in terms of the various moduli of continuity. In the case of the heat equation we have the following result on the behavior of u near a jump discontinuity of ϕ:

THEOREM 2. Let

$$L(\xi) = \sum_{k=1}^{n} \xi_k^2$$

and let $u = T_t \phi$ be real. Let $E(a)$ = the set of x such that $\phi(x) \leq a$, and let

$$F_{x,r}(a) = \frac{\text{measure } (E(a) \cap S(x,r))}{\text{measure } (S(x,r))}$$

where $S(x,r)$ is the sphere with center at x and radius r. Suppose that

$$\lim_{r \to 0} F_{x,r}(a) = F(a)$$

at all points of continuity of F, where F is a probability distribution function with finite first moment. Then

$$\lim_{t \to 0+} u(x,t) = \int_{-\infty}^{\infty} a\, dF(a) = \text{math. expectation of } F$$

If $n = 1$ and ϕ has a jump discontinuity at x_0, then $u(x,t) \longrightarrow (\phi(x_0+0) + \phi(x_0-0))/2$ as $(x,t) \longrightarrow (x_0,0)$ along any path in the (x,t)-plane, not tangential to the x-axis at $(x_0,0)$. If the path C is tangential to the x-axis at $(x_0,0)$ but has a curvature there, then $\lim u(x,t)$ as $(x,t) \longrightarrow (x_0,0)$ along C is a weighted mean of $\phi(x_0+0)$ and $\phi(x_0-0)$, the weights depending on the curvature of C at $(x_0,0)$ and being determined by the Gaussian distribution function.

These results are basic to the uniqueness and representation theorems for (1).

THEOREM 3. If u satisfies (1) and

$$\int_{E_n} |u(x,t)| \exp\left\{-M \|x\|^{\mu}\right\} dV_x \leqq N \tag{7}$$

for $0 < t < c$, if u is continuous for $0 \leqq t < c$, and if $u(x,0) \equiv 0$, then $u \equiv 0$. If u satisfies (7) for $0 < t < c$ where $c \leqq (C_3/M)^{m-1}$, and is continuous for $0 \leqq t < c$, $u(x,0) = \phi(x)$, then $u = T_t\phi$ for $0 < t < c$.

This is a sharpening of a result of Ladyzhenskaya [5], obtained by combining her methods with those of [11]. Probably μ cannot be replaced by any larger number, but the conditions on the manner in which (x,t) approaches $\phi(x)$ as $t \longrightarrow 0$ can probably be relaxed considerably. For example, if $|u(x,t)| \leqq g(x)$, $0 < t < c \leqq (C_3/M)^{m-1}$, and $g(x) \exp\left\{-M \|x\|^{\mu}\right\} \in L_1(E_n)$, and

$$\lim_{t \to 0} u(x,t) = \phi(x)$$

almost everywhere, then certainly $u = T_t\phi$.

In the case of the heat equation sharper results are known.

THEOREM 4. Let

$$L(\xi) = \sum_{k=1}^{n} \xi_k^2$$

and let u be a solution of (1) for $0 < t < c \leq 1/(4M)$. Then for u to be representable as a Poisson-Stieltjes integral, it is necessary and sufficient that u satisfy (7). (Note: here $\mu = 2$). If

$$\lim_{t \to 0} u(x,t) = 0$$

for all x, then $u \equiv 0$.

This theorem was obtained in 1946 (see [11]) and for the case $n = 1$ was rediscovered independently by Fulks [3] in 1949. It generalizes and sharpens results of Tykhonov [16], Widder [19] and Pollard [10]. An analogous result for $n = 1$ and a finite rectangle was obtained independently by Fulks [3] and by Hartman and Wintner [4]. A similar theorem holds for the case of the Laplace equation in any number of dimensions for spherical domains. In this connection Lohwater [6] has obtained a result of overlapping generality. The only proof known to the author depends on the deep results of Besicovich [1] (see also Morse [8]). For the case $n = 1$ we know several proofs, none of which is elementary. The essential lemma is:

A completely additive set function, whose upper symmetric derivate is everywhere non-negative, is non-negative. This sharpens a theorem of Ward [18]. Note that in the uniqueness theorem 4 above, the phrase "for all x" cannot be replaced by "almost everywhere."

While the growth condition (7) is necessary and sufficient for the representation by a Poisson-Stieltjes integral, it is stronger than necessary for the uniqueness theorem. Growth conditions which are essentially the best possible were obtained by Täcklind [13]. For the heat equation with $n = 1$ we can improve Täcklind's result slightly.

THEOREM 5. Let H be a measurable function on E_1, let A be an arbitrary positive constant, let $g(R,M)$ = the measure of the set of x such that $R \leqq x \leqq R + A$, $H(x) \leqq M$, where $R \geqq 0$,

$$L(R) = \inf \left(\frac{M - \log g(s,M)}{s}\right) \qquad \text{for } s \geqq k, \quad -\infty < M < +\infty$$

If

$$(*) \qquad \frac{\partial^2 u}{\partial x^2} = \frac{\partial u}{\partial t}$$

and

$$\int_{-\infty}^{\infty} |u(x,t)|\, e^{-H(|x|)}\, dx \leqq M \quad \text{for} \quad 0 < t < c,$$

and

$$\lim_{t \to 0+} u(x,t) = 0 \quad \text{for all } x,$$

and

$$\int_1 \frac{dR}{L(R)} = +\infty ,$$

then $u \equiv 0$.

If

$$\int_1^{\infty} \frac{dR}{L(R)} < +\infty$$

and $0 < \alpha < 1$, then there is a solution of (*) in some strip $0 < t < c$ and an M such that

(a) $\displaystyle\int_{-\infty}^{\infty} |u(x,t)|\, e^{-\alpha H(|x|)}\, dx \leqq M < +\infty$ for $0 < t < c$,

(b) $\displaystyle\lim_{t \to 0+} u(x,t) = 0$ for all x

(c) $u \equiv 0$

If $H(|x|) \geq h > -\infty$ for almost all x, then we can take $\alpha = 1$.

By oral communication from Beurling, the above gap between the necessary and sufficient conditions for uniqueness can be eliminated and the conditions themselves can be considerably simplified.

We remark that for solutions of the heat equation satisfying (7) there are theorems analogous to those of Fatou and Phraegmen-Lindelöf for harmonic and analytic functions. The following isolated and curious result may be noted: if u is a solution of the heat equation which satisfies (7), then $\sqrt{t}\, u(x,t)$ is of bounded variation for each fixed x, and

$$\lim_{t \to 0} \sqrt{t}\ u(x,t) = g(x)$$

exists and is finite for all x. Of course $g(x) = 0$ almost everywhere.

We now turn to the non-homogeneous problem

$$\text{(8)} \qquad \frac{\partial u}{\partial t} - L\left(\frac{\partial}{\partial x}\right)u = f(x,t), \qquad u(x,0) = \phi(x)$$

The formal solution is

$$u = T_t\phi + \int_0^t T_{t-\tau} f(x,\tau)\, d\tau = u_1 + u_2$$

If

$$\int_{E_n} |f(x,t)| \exp\left\{-M \|x\|^{\mu}\right\} dV_x \leq N < +\infty$$

for $0 \leq t < c$, then

$$\lim_{t \to 0} u_2(x,t) = 0$$

almost everywhere. If besides f satisfies a Hölder condition in every bounded domain, then u_2 and all its partial derivatives of order $\leq m$ with respect to $x_1, \ldots, x_n$ exist and are continuous for $t > 0$ and u satisfies (8).

Let $L_s(\xi) = L(\xi) - s$, where s is an arbitrary complex number. The corresponding fundamental solution is

$$K_s(x,t) = e^{-st} K(x,t)$$

If $f(x) \in L_1(E_n)$ and $R(s) > C_2$, then

$$(Pf)\ (x) = \int_0^\infty e^{-s\tau}(T_\tau f)\ (x)d\tau$$

exists for all x. Consider the problem

(9) $$\frac{\partial u}{\partial t} - L_s\left(\frac{\partial}{\partial x}\right) u = f(x), \qquad u(x,0) = \phi(x)$$

If $f \in L_1(E_n)$ and $\phi(x) \exp\left\{- \epsilon \ \|x\|^\mu\right\} \in L_1(E_n)$ for all $\epsilon > 0$, then u can be represented in the form $u_1 + u_2$ above for all $t > 0$. If $R(s) > C_2$, then $u_1 \longrightarrow 0$, $u_2 \longrightarrow Pf$ uniformly as $t \longrightarrow +\infty$, so that $u \longrightarrow Pf$. Pf is the unique solution of

(10) $$L_s u = -f$$

such that $u \in L_1(E_n)$. This existence and uniqueness theorem for solutions of (10) with $R(s) > C_2$ may be new. Tykhonov [17] has studied the approach of solutions of the heat equation to a steady state, and Bruk (see Petrovskiĭ [9]) for some more general equations. See also Milgram and Rosenbloom [7].

BIBLIOGRAPHY

[1] BESICOVITCH, A. S., "A general form of the covering principle and relative differentiation of additive functions." Proceedings of the Cambridge Philosophical Society, 41 (1945), 103-110; 42 (1946), 1-10.

[2] BOCHNER, S. and CHANDRASEKHARAN, K., Fourier Transforms, Princeton University Press, Princeton (1949).

[3] FULKS, W. B., Thesis, University of Minnesota, 1949.

[4] HARTMAN, P. and WINTNER, A., "On the solutions to the equation of heat conduction," American Journal of Mathematics, 72 (1950), 367-395.

[5] LADYZHENSKAYA, O. A., "On the uniqueness of the solution of the Cauchy problem for a linear parabolic equation," Matematiceskii Sbornik, N.S. 27, 69 (1950), 175-184.

[6] LOHWATER, A. J., "A uniqueness theorem for a class of harmonic functions," Proceedings of the American Mathematical Society, 3 (1952), 278-279.

[7] MILGRAM, A. N. and ROSENBLOOM, P. C., "Harmonic forms and heat conduction. I. Closed Riemann manifolds," Proceedings of the National Academy of Sciences of the U. S. A., 37 (1951), 180-184.

[8] MORSE, A. P., "Perfect blankets," Transactions of the American Mathematical Society, 61, (1947), 418-442.

[9] PETROVSKIĬ, I. G., "On some problems of the theory of partial differential equations," Uspekhi Matematischeskikh Nauk (N.S.), 1, No. 3-4 (13-14), (1946), 44-70. (American Mathematical Society Translation No. 12).

[10] POLLARD, H., "One-sided boundedness as a condition for the unique solution of certain heat equations," Duke Mathematical Journal 11, (1944), 651-653.

[11] ROSENBLOOM, P. C., "Studies on the heat equation," (Abstract). Bulletin of the American Mathematical Society, 53, (1947), 56.

[12] ROSENBLOOM, P. C., "Lectures on partial differential equations," University of Minnesota (1951). In preparation.

[13] TÄCKLIND, S., "Sur les classes quasianalytiques des solutions des équations aux dérivées partielles du type parabolique." Thesis, Uppsala (1936).

[14] TITCHMARSH, E. C., Introduction to the Theory of Fourier Integrals, Oxford University Press, 1937.

[15] TRJITZINSKY, W. J., "Analytic theory of linear differential equations," Acta Mathematica, 62 (1934), 167-226.

[16] TYKHONOV, A., "Théorèmes d'unicité pour l'équation de la chaleur," Adakemiya Nauk USSR Matematiceckii Sbornik, 42, (1935), 199-215.

[17] TYKHONOV, A., "Sur l'équation de la chaleur à plusieurs variables." Moscow. Universitet. Bulletin, Série internationale, Section A, Mathématiques et Mécanique, 1, No. 9, (1938), 1-44.

[18] WARD, A. J., "On the derivation of additive functions of intervals in m-dimensional space," Fundamenta Mathematicae, 28 (1937), 265-279.

[19] WIDDER, D. V., "Positive temperatures on an infinite rod." Transactions of the American Mathematical Society, 55, (1944), 85-94.

XI. ON LINEAR HYPERBOLIC DIFFERENTIAL EQUATIONS WITH VARIABLE COEFFICIENTS ON A VECTOR SPACE

J. Leray

Abstract: In §1 of this paper a mistake in Petrowsky's proof of his fundamental existence theorem is pointed out; §2 shows how this theorem can be proved and §3 shows how it holds for merely Lipschitz continuous coefficients. This theorem is extended to systems and manifolds in another paper [2].

§1. INTRODUCTION

Cauchy and Kowalewski gave a local existence theorem for analytic equations. J. Hadamard [1] pointed out that for physical problems such a local theorem is useless and proved a global existence theorem for hyperbolic equations of second order; he used Riemannian geometry and Green's method which transforms the problem of finding regular solutions into a problem of finding solutions having a given singularity. Later on J. Schauder [4] gave an easier method: the classical energy relation enables him to extend the local Cauchy-Kowalewski solution into a global one for analytic and finally non-analytic equations of second order. Two years later I. Petrowsky [3] gave the definition of hyperbolic equations of any order and, using Schauder's process, proved a global existence theorem for these equations.

The main point of Petrowsky's paper is to find an a priori bound for a solution of Cauchy's problem for equations in l independent variables; he obtains his bound after 17 pages of inequalities without comments. His first step is to define a strange transformation involving both the Fourier transformation and the use in the $(l-1)$-dimensional space of a variable frame that depends on its first vector. He assume (pp. 821-822 and more explicitly in the last lines of p. 861) that this frame depends continuously on its first vector. But this assumption does not differ from the assumption that the $(l-2)$-sphere is parallelisable; i.e., that there exists on its surface a frame which depends continuously on its vertex. It is a known topological theorem that even-dimensional spheres cannot have any continuous vector field and therefore are not parallelisable.

In fact, Steenrod and Whitehead [5] have recently proved a deep theorem which asserts that a sphere whose dimension is not 2^k-1 is not parallelisable. If one could prove that there exist parallelisable spheres with arbitrarily large dimensions, Hadamard's descent method would enable us to complete Petrowsky's proof but at present we know only that the 1-, 3- and 7-dimensional spheres can be parallelised by means of complex numbers, quaternions and Cayley numbers [6]. We conclude: Petrowsky has actually established his a priori bound only when the number l of independent variables in the equations is $l \leq 9$. Let us outline in the next section how such a priori bounds can be obtained for any value of l by a method much simpler than Petrowsky's.

§2. EQUATIONS WITH INFINITELY OFTEN DIFFERENTIABLE COEFFICIENTS

Let X be an l-dimensional vector space and Σ its dual; consider a linear differential equation of order m

$$a(x,p)u(x) = v(x) \tag{1}$$

where

$$x = (x_1, x_2, \ldots, x_l) \in X, \quad p = (p_1, p_2, \ldots, p_l) = \left(\frac{\partial}{\partial x_1}, \frac{\partial}{\partial x_2}, \ldots, \frac{\partial}{\partial x_l}\right) \in \Sigma$$

Suppose Cauchy data $u, p_1 u, \ldots, p_1^{m-1} u$ are given on the hyperplane

$$x_1 = \text{constant}$$

we seek an a priori bound for the solution of Cauchy's problem.

Replace equation (1) by a first order system, introducing new unknowns for the first $m-1$ derivatives of u with respect to x_1:

$$\begin{cases} p_1 u_1 = u_2, \quad p_1 u_2 = u_3, \ldots, \quad p_1 u_{m-1} = u_m \\ p_1 u_m = -c_0(x,p)u_1 - c_1(x,p)u_2 - \cdots - c_{m-1}(x,p)u_m + v \end{cases}$$

here

$$u_1(x) = u(x)$$

$$a(x,p) = p_1^m + c_{m-1}(x,p)p_1^{m-1} + \cdots + c_1(x,p)p_1 + c_0(x,p)$$

and $c_\lambda(x,p)$ is independent of p_1.

We write the system in the matrix form

$$p_1U(x) = A(x,p)U(x) + V(x) \tag{2}$$

where

$$U = (u_1,u_2,\dots,u_m), \quad V = (v_1,v_2,\dots,v_m)$$

and

$$A = \begin{pmatrix} 0 & 1 & 0 & . & 0 \\ 0 & 0 & 1 & . & 0 \\ 0 & 0 & 0 & . & 0 \\ . & . & . & . & . \\ 0 & 0 & 0 & . & 1 \\ -c_0 & -c_1 & -c_2 & . & -c_{m-1} \end{pmatrix} \tag{3}$$

thus the elements $a_{\lambda\mu}(x,p)$ of $A(x,p)$ are independent of p_1 and have (at most) the order $\lambda - \mu + 1$.

Now suppose we know another real matrix $B(x,p)$ with the following properties:

(a) The rank of B is the same as that of A.

(b) The elements $b_{\lambda\mu}(x,p)$ of B are independent of p_1; they have the order $2n - \lambda - \mu$. Hence, the $(\lambda,\mu)^{th}$ element of the product matrix BA has (at most) the order $2n - \lambda - \mu + 1$.

(c) However, the $(\lambda,\mu)^{th}$ element of the matrix $BA + (BA)^*$ has the smaller order $2n - \lambda - \mu$ (C^* denotes the adjoint of C).

(d) $B(x,p)$ is a hermitian positive definite operator, i.e.,

$$B = B^*$$

$$(B(x,p)U,U)_t > k \sum_{\substack{1 \leq \lambda \leq l \\ 0 \leq \mu \leq m}} \| p^{m-\mu} u_\mu(x) \|_t^2$$

where k is a positive constant and our inner product is defined as follows:

$$(u,v)_t = \int \cdots \int_{x_1 = t} u(x)v(x)\, dx_2 dx_3 \dots dx_l$$

$$(U,V)_t = \sum_{1 \leq \lambda \leq l} (u_\lambda, v_\lambda)_t$$

$$\|u\|_t^2 = (u,u)_t$$

If $B(x,p)$ has the above properties, we say that the hermitian part of $A(x,p)$ is bounded in the sense of the norm induced by $B(x,p)$. Indeed, for the norm $\sqrt{(BU,U)_t}$, the adjoint of A is $B^{-1}A^*B$ and the hermitian part of A is

$$\frac{1}{2} A + \frac{1}{2} B^{-1}A^*B$$

which is bounded since $((BA + A^*B)U,U)_t < \text{const.}(BU, U)_t$.

Whenever the hermitian part of A is bounded in the norm induced by B, an "energy inequality" holds: From equation (2),

$$\frac{d}{dt}(BU,U)_t = ((BA + A^*B)U,U)_t + (B_tU,U)_t + 2(BU,V)$$

$$< \text{const.}\ (BU,U)_t + \text{const.}\ \sqrt{(BU,U)_t(BV,V)_t}$$

Let γ be a positive constant. Then

$$\frac{d}{dt}\left[\sqrt{(BU,U)_t}\ e^{-\gamma t}\right] < \text{const.}\ \sqrt{(BV,V)_t}\ e^{-\gamma t} \tag{4}$$

If we integrate (4) we obtain a priori bounds for $u(x)$ in terms of the norm induced by B. (If n can be chosen arbitrarily large, then arbitrarily high derivatives of u are involved in this norm.)

We shall call a real matrix A symmetric with respect to the norm induced by a real, symmetric, positive definite matrix B if the matrix BA is symmetric. A simple application of Fourier transforms yields operators $B(x,p)$ with respect to which the hermitian part of $A(x,p)$ is bounded (and hence, yields a priori bounds for the solution of our Cauchy problem) provided we can solve the following purely algebraic problem:

Let $H(x,\xi)$ be the matrix which we obtain when, in the matrix $A(x,\xi)$, we replace the elements $c_\lambda(x,\xi)$ by their principal parts $\tilde{c}_\lambda(x,\xi)$; then the elements $h_{\lambda\mu}(x,\xi)$ of $H(x,\xi)$ are polynomials in $(\xi_2,\ldots,\xi_l)$, homogeneous of degree $\lambda - \mu + 1$. We ask for a matrix $B(x,\xi)$ such that $H(x,\xi)$ is symmetric with respect to the norm induced by $B(x,\xi)$:

B is a symmetric, positive definite matrix, BH is symmetric; the elements $b_{\lambda\mu}(x,\xi)$ of $B(x,\xi)$ have to be polynomials in $(\xi_2,\xi_3,\ldots,\xi_l)$ homogeneous of degree $2n - \lambda - \mu$.

This problem can be solved only if all characteristic roots of $H(x,\xi)$ are real. When they are real and distinct, the problem has solutions even if A is not of the special type (3); but if A is of type (3) then the solution is particularly simple as we shall now show.

Let $P(t)$ be the characteristic polynomial of $H(x,\xi)$:

$$P(t) = t^m + \tilde{c}_{m-1}(x,\xi)t^{m-1} + \dots + \tilde{c}_1(x,\xi)t + \tilde{c}_0(x,\xi)$$

The roots $t_1, t_2, \dots, t_m$ of $P(t)$ are assumed to be real and distinct. Let

$$s_\lambda = \sum_{\mu=1}^{m} t_\mu^\lambda$$

and

$$T = \begin{pmatrix} 1 & 1 & 1 & . & 1 \\ t_1 & t_2 & t_3 & . & t_m \\ t_1^2 & t_2^2 & t_3^2 & . & t_m^2 \\ \dots & \dots & \dots & \dots & \dots \\ t_1^{m-1} & t_2^{m-1} & t_3^{m-1} & . & t_m^{m-1} \end{pmatrix}$$

On the one hand, the matrix

$$S = \begin{pmatrix} s_0 & s_1 & s_2 & . & s_{m-1} \\ s_1 & s_2 & s_3 & . & s_m \\ s_2 & s_3 & s_4 & . & s_{m+1} \\ \dots & \dots & \dots & \dots & \dots \\ s_{m-1} & s_m & s_{m+1} & . & s_{2m-2} \end{pmatrix} = TT'$$

(where T' is the transpose of T) is symmetric and positive definite. On the other hand, the s_λ are polynomials in the $\tilde{c}_\lambda(x,\xi)$ and can be obtained by a classical induction formula so that we can easily verify that

$$HS = \begin{pmatrix} s_1 & s_2 & s_3 & . & s_m \\ s_2 & s_3 & s_4 & . & s_{m+1} \\ s_3 & s_4 & s_5 & . & s_{m+2} \\ \dots & \dots & \dots & \dots & \dots \\ s_m & s_{m+1} & s_{m+2} & . & s_{2m-1} \end{pmatrix}$$

is symmetric. Therefore, our desired matrix can be chosen as

$$B(x,\xi) = (\xi_2^2 + \xi_3^2 + \ldots + \xi_1^2)^{n - \frac{m(m-1)}{2} - 1} |\det S| \cdot S^{-1}$$

Now that we have obtained a priori bounds for the solution of our Cauchy problem, we can, by means of Schauder's process, extend the local Cauchy-Kowalewski solution to a global one and thus obtain the existence theorem; then, using the well-known method of Holmgren, we obtain the uniqueness theorem.

The assumption that the roots of $P(t)$ are real and distinct is equivalent to the assumption that $a(x,p)$ is <u>hyperbolic</u> according to Definition 1, and to the assumption that the first axis of Σ lies in the interior of the director cone $\Gamma_x(a)$ defined below.

DEFINITION 1. Let $a(x,\xi)$ be a polynomial of degree m, and $h(x,\xi)$ the sum of its homogeneous terms of degree m. At a given point x, $h(x,\xi) = 0$ defines a cone in the space Σ.

$a(x,p)$ is said to be <u>hyperbolic</u> at the point x if there exists at least one point ξ_0 in Σ such that any real line through ξ_0 and not through the origin cuts the cone $h(x,\xi) = 0$ at m real and distinct points. All elements ξ satisfying this condition at a point x form a double convex cone $\Gamma_x(a)$, $-\Gamma_x(a)$ whose boundary belongs to the cone $h(x,\xi) = 0$; $h(x,\xi) = 0$ has no singular generators.

We can now draw the following conclusion:

PROPOSITION 1. Assume that $a(x,p)$ is hyperbolic everywhere in X, that its coefficients are infinitely often differentiable and that the interior of the set $\Gamma_X(a) = \bigcap_{x \in X} \Gamma_x(a)$ is not empty. Assume that $v(x)$ is infinitely often differentiable and vanishes outside a bounded region. Then equation (1) has a unique solution

$$u(x) = a(x,p)^{-1}\, v(x)$$

such that $u(x) \exp.(-x \cdot \xi)$ is square integrable on X for any ξ belonging to some domain whose director cone is $\Gamma_X(a)$.

REMARK 1. Let $u'(x)$ be a derivative of any order of $u(x)$; $u'(x) \exp. (-x\cdot\xi)$ is square integrable on X for any ξ belonging to some domain whose director cone is $\Gamma_X(a)$.

REMARK 2. Let $S(u)$ be the support of u (i.e., the smallest closed subset of X outside which $u = 0$); let $C_X(a)$ be the cone dual to $\Gamma_X(a)$. Then, by Proposition 1,

$$S(u) \subset S(v) + C_X(a) \tag{5}$$

Thus Proposition 1 solves the Cauchy problem for data zero at infinity in the direction $-C_X(a)$. Any "well posed" Cauchy problem can now be reduced to this one.

REMARK 3. The integral of the square of $u(x) \exp. (-x\cdot\xi)$ can be bounded by the integral of the square of $v(x) \exp. (-x\cdot\xi)$. However, such a bound cannot be obtained from the inequality (4) and the matrix $B(x,\xi)$.

§3. EQUATIONS WITH LIPSCHITZ-CONTINUOUS COEFFICIENTS

We shall now obtain new a priori bounds by a simpler process which does not employ any special choice of coordinates, but which supposes that the solution $u(x)$ of equation (1) is defined in the whole space X. Evidently such a priori bounds could not be the basis of Schauder's method used in the proof of Proposition 1; it dealt with solutions defined only in strips.

Let

$$P(\lambda) = \lambda^m + \dots \qquad Q(\lambda) = \lambda^{m-1} + \dots$$

be two real polynomials in one variable λ. We say that the roots of $Q(\lambda)$ separate the roots of $P(\lambda)$ if

(a) all the roots of both polynomials are real and distinct,

(b) letting $\lambda_1 < \lambda_2 < \dots < \lambda_m$ denote the roots of $P(\lambda)$ and $\mu_1 < \mu_2 < \dots < \mu_{m-1}$ the roots of $Q(\lambda)$,

$$\lambda_1 < \mu_1 < \lambda_2 < \mu_2 < \dots < \mu_{m-1} < \lambda_m$$

If the roots of $Q(\lambda)$ separate the roots of $P(\lambda)$, then the quotient

(6) $$P(\lambda)/Q(\lambda)$$

maps the half plane $\mathrm{Im}(\lambda) > 0$ onto itself. (This property of (6) is classic.)

Now let $a(\xi)$ and $b(\xi)$ be two homogeneous hyperbolic polynomials of degrees m and $m-1$; we say that the sheets of the cone $b(\xi) = 0$ separate the sheets of the cone $a(\xi) = 0$ if any line not through the origin with direction in $\Gamma_X(a)$ cuts the cone $b(\xi) = 0$ at points which separate the points where that same line cuts the cone $a(\xi) = 0$. If this is the case we have

$$\mathrm{Re}[b(\bar{\zeta})a(\zeta)] \neq 0$$

where

$$\zeta = \xi + i\eta, \quad \bar{\zeta} = \xi - i\eta, \quad \xi \text{ in } \Gamma_X(a), \quad \eta \text{ in } \Sigma$$

If $a(\xi)$ is given we can choose for instance

$$b(\xi) = \xi_1^* \frac{\partial a(\xi)}{\partial \xi_1} + \dots + \xi_l^* \frac{\partial a(\xi)}{\partial \xi_l}$$

where ξ^* is in $\Gamma_X(a)$.

Using Laplace transforms, we arrive at the following assertion: With any hyperbolic operator $a(x,p)$ of order m we can associate hyperbolic operators $b(x,p)$ of order $m-1$ such that, for the Hilbert norm

(7) $$\|f\| = \left[\int_X [f(x)\exp(-x\cdot\xi)]^2\, dx\right]^{1/2} \quad (\xi \text{ in } \Gamma_X(a), \ \|\xi\| \text{ large})$$

the hermitian operator of order $2(m-1)$

$$[b(x,p)]^* a(x,p) + [a(x,p)]^* b(x,p)$$

is positive definite; hence bounds for $a(x,p)^{-1}$ and $a(p,x)^{-1}$ can be obtained and enable us to extend Proposition 1 as follows:

PROPOSITION 2. Assume that the coefficients of $a(x,p)$ or $a(p,x)$ are Lipschitz continuous. Then

(a) $a(x,p)^{-1}v(x)$ (and its derivatives) of order $< m$ (of order $\leq m$) are locally square integrable if $v(x)$ (and its first derivatives) are locally square integrable.

(b) $a(p,x)^{-1}v_X$ (and its first derivatives) are locally square integrable if v_X is a derivative of order $\leq m-1$ (of order $< m-1$) of a square integrable function $v(x)$ whose support is bounded.

Now we can easily extend Propositions 1 and 2 to <u>hyperbolic systems</u> and also to <u>manifolds</u> and state a more precise relation between $S(u)$ and $S(v)$ than (5), cf. [2].

REMARK 4. We can now express the fundamental inequality as follows, using $a(x,p)^{-1}$: if, in the space Σ, the sheets of the cone $b(x,\xi) = 0$ separate the sheets of the cone $a(x,\xi) = 0$, if ξ is in the interior of $\Gamma_X(a)$, if $\|\xi\|$ is large and if

$$a(x,\xi)b(x,\xi) > 0$$

then the norms

$$\|\xi\| \sum_{\substack{1\leq\lambda\leq l\\ 0\leq\mu\leq m}} \| p_\lambda^\mu\, a(x,p)^{-1}v \|$$

(where $\| \;\|$ is defined by (7)) and

$$(v(x), b(x,p)a(x,p)^{-1}\, v(x))^{1/2}$$

are equivalent. This is not much more than the following assertion: the hermitian part of $b(x,p)a(x,p)^{-1}$ is positive definite. This last assertion is merely an extension of the classic property of (6).

BIBLIOGRAPHY

[1] HADAMARD, J., Le Problème de Cauchy et les équations aux dérivées partielles linéaires hyperboliques, Hermann et Cie, Paris (1932).

[2] LERAY, J., "The linear hyperbolic differential equation," (to be published in) Bulletin of the American Mathematical Society; cf. also mimeographed lecture notes at the Institute for Advanced Study, Princeton (1952).

[3] PETROWSKY, I., "Über das Cauchysche Problem für Systeme von partiellen Differentialgleichungen," Recueil math. (Mat. Sbornik) 2 44 (1937), 815-868.

[4] SCHAUDER, J., "Das Anfangswertproblem einer quasilinearen hyperbolischen Differentialgleichung zweiter Ordnung in beliebiger Anzahl von unabhängigen Veränderlichen," Fundamenta Mathematicae 24 (1935), 213-246.

[5] STEENROD, N. E., WHITEHEAD, J. H. C., "Vector fields on the n-sphere," Proceedings of the National Academy of Sciences 37 (1951), 58.

[6] STIEFEL, E., "Richtungsfelder und Fernparallelismus in n-dimensionalen Mannigfaltigkeiten,"Commentarii Mathematici Helvetici 8 (1935), 347.

XII. THE INITIAL VALUE PROBLEM FOR NONLINEAR HYPERBOLIC EQUATIONS IN TWO INDEPENDENT VARIABLES[1]

P. D. Lax

INTRODUCTION

This article gives an account of recent developments in the theory of the initial value problem for hyperbolic equations in two independent variables. The objects of our attention are first order systems:

$$U_t + AU_x + B = 0 \tag{1}$$

U is a column vector of n unknown functions. If the coefficient matrices A and B depend on x,t and U, the system is called quasi-linear, if A is independent of U, the system is called semi-linear, and if in addition B is a linear function of U, the system is called linear.

If all eigenvalues of A are real, and if A has a full set of linearly independent eigenvectors at all points of a certain domain of x,t,U space, the system (1) is called hyperbolic in this domain; the eigenvalues and eigenvectors are assumed to be at least once continuously differentiable functions of x,t,U.

The initial value problem is to find a solution U of the system (1) whose value is prescribed at $t = 0$ along some interval of the x axis; in this paper the initial interval will be the whole x axis, so that we may prescribe

$$U(x,0) = \phi(x), \quad -\infty < x < \infty \tag{2}$$

The domain of definition of solutions will be a strip,

[1] Some of the results announced in this paper are new and obtained under contract with the Office of Naval Research; some of those in Section 3 were obtained for the Los Alamos Scientific Laboratory. I am grateful to both organizations for their support; in addition, my thanks are due to members of the staff of the Los Alamos Scientific Laboratory for their help in carrying out numerical computations; some of these were done on IBM equipment, but mostly on the Los Alamos MANIAC and the Remington Rand Eckert-Mauchly UNIVAC.

$$0 \leq t \leq \delta, \quad -\infty < x < \infty$$

The initial value problem is correctly set for hyperbolic equations; i.e., the initial value problem has (under some differentiability assumptions) a unique solution which shares whatever differentiability properties the initial vector might have, and the solution depends - in terms of a variety of norms - continuously on the initial data. This is true both for linear and nonlinear equations, with this important difference: The solution of an initial value problem for a linear system exists for all times, while in the nonlinear case it may exist only in the small; i.e., the solution may after a finite time become nondifferentiable, and be incapable of being continued as a regular solution. There is nothing one can do about this in general, but we have a strong reason to believe that for a certain class of problems, e.g., the initial value problems of hydrodynamics which describe the unfolding of a physical system from its initial state, the solutions must exist, in some generalized sense, for all time. The basic concepts of such a theory in the large are sketched Section 3.

The theory in the small is discussed in Sections 1 and 2. Our attitude is this: Solving the initial value problem for a given system of equations induces a one-parameter family of transformations $S(t)$ which assign to the initial value of a solution of (1) its value at t. The aim of the theory is to study properties of this family of transformations, in particular those properties which have an aesthetic appeal or which have significance for physical applications.

Initial value problems for <u>higher order</u> quasilinear hyperbolic systems can be reduced to initial value problems for first order hyperbolic systems by introducing as new unknowns all partial derivatives of order less than the highest. The results presented in this survey include - via this reduction - many of the existing results on the initial value problems for higher order systems; however, more refined results, as in Part I of P. Hartman and A. Wintner [14], cannot be obtained this way. Almost all theorems known on the initial value problem for <u>general nonlinear</u> hyperbolic equations can be deduced from results for <u>quasi-linear</u> systems by differentiating the nonlinear equation with respect to t, thus making a quasilinear system out of it. The only result on nonlinear hyperbolic equations known to me which cannot be obtained by such a reduction is the Haar uniqueness theorem (see [11]) and the theory of the initial value problem of the Monge-Ampère equation (see Lewy [23] Courant-Hilbert [3], pp. 344-345, or [14]), which plays such an important role in Lewy's theory of the elliptic Monge-Ampère equation.

Before presenting the theory of quasilinear hyperbolic equations, a few remarks about the development of the subject are in order:

The first existence theorem for the solution of an initial value problem is due to E. Picard who dealt with a single second order semilinear equation. This existence theorem was extended by O. Perron, [26], to arbitrary first order semilinear systems. The linear case was treated previously by E. Holmgren, [15].

H. Lewy, in [21], proved that nonlinear second order hyperbolic equations have solutions with prescribed initial data; an extension of this result by him and Friedrichs, [6], is not complete. The first complete existence theorem for the general quasilinear case is due to J. Schauder [28]; subsequent proofs were given by M. Cinquini-Cibrario [1], K. O. Friedrichs [9], R. Courant and P. D. Lax [4], A. Douglis [5], P. Hartman and A. Wintner [14] and P. D. Lax [18].

§1. THEORY OF GENUINE (DIFFERENTIABLE OR LIPSCHITZ CONTINUOUS) SOLUTIONS

One of the basic tools for studying the initial value problem is an a priori estimate due to Haar [12], see also [28] (already used implicitly by previous writers). This estimate, expressed in words, states that the solution of a linear hyperbolic system depends boundedly on the initial values and on the inhomogeneous term. The norm in which boundedness is to be measured is the maximum norm; the magnitude of the bound depends on upper bounds for B, A and first derivatives of A.

That the solution is a bounded function of the initial data in the sense of the maximum norm is a special feature of linear hyperbolic equations in one space variable, which is no longer true if the number of space variables is greater than one. In order to preserve this feature of boundedness for more space variables one has to resort to another norm, in terms of which the dependence of the solution on the initial data is bounded; the L_2 norm was used by K. O. Friedrichs and H. Lewy [7] for second order hyperbolic systems, by K. O. Friedrichs [8] for symmetric hyperbolic systems, and by I. Petrovsky [27] for general hyperbolic systems. (A new analysis of the general hyperbolic case has been given recently by J. Leray, see [20]).

We shall first state the Haar estimate in the special case where the matrix A is a diagonal matrix D. Let the equation be

$$U_t + DU_x + BU + C = 0 \tag{1.1}$$

with initial values $U(x,0) = \Phi(x)$.

HAAR'S INEQUALITY: Denote by $\phi_0, \beta, C(\tau)$ upper bounds for $|\Phi(x)|$, $|B(x,t)|$, $|C(x,\tau)|$ for all values of x and t.[2] Then

$$(1.2) \qquad |U(x,t)| \leq \phi_0 e^{\beta t} + \int_0^t C(\tau) e^{\beta(t-\tau)} d\tau$$

Each equation in the system (1.1) may be considered an ordinary differential equation along a characteristic. The inequality (1.2) can be derived by the method used for obtaining the corresponding classical estimate for solutions of ordinary differential equations. Note that (1.2) does indeed mean that U is a bounded function of ϕ and C; the magnitude of the bound depends only on the magnitude of $|B|$.

Any linear (or semilinear) hyperbolic system can be put in diagonal form by introducing new independent variables V related to the old by $U = TV$, where T is a matrix such that $T^{-1}AT = D$ is diagonal. Such a (real) matrix T exists since A was assumed to have a full set of linearly independent eigenvectors. V satisfies the equation

$$(1.3) \qquad V_t + DV_x + \tilde{B}V + \tilde{C} = 0$$

where $\tilde{B}$ and $\tilde{C}$ involve the coefficients A, B, C, of (1.1), the matrices T, T^{-1} and the first derivatives of T. Since we know these relationships and since we have a Haar estimate for a system of the form (1.3), we also have an estimate for a system of the form (1.1).

If $\tilde{B}$ happens to be zero, each equation in (1.3) can be integrated along the corresponding characteristic starting on the initial interval; thus if the initial values of V are known, V can be determined uniquely for all time to come. If $\tilde{B}$ is not zero, the solution can be obtained by iterating the transformation $V = \tau(W)$, defined for all once differentiable W by the equation

$$(1.4) \qquad V_t + DV_x + \tilde{B}W + \tilde{C} = 0, \qquad V(x,0) = \phi(x)$$

On the basis of the Haar inequality one can show that iterates of this transformation converge to a solution of the initial value problem; this solution has as many derivatives as desired if the coefficients $D, \tilde{B}, \tilde{C}$ of (1.3) and the initial vector are sufficiently differentiable. There is no difficulty in extending this argument to the semilinear case. What is less obvious is that solutions of quasilinear equations also can be constructed by such an iteration; it turns out that they can, provided that

---

[2] The norm of a vector is the largest of the absolute values of its components; the norm of a matrix is its norm as a transformation.

one defines the domain of the transformation τ carefully. $U = \tau(W)$ is now defined as the solution of

(1.5) $$U_t + A(W)U_x + B(W) = 0, \quad U(x,0) = \phi(x)$$

As a precautionary measure assume that A, B, ϕ, W are sufficiently differentiable functions so that the equation (1.5) has a solution U, which is at least three times differentiable.

Denote by W_0 the function $W_0(x,t) = \phi(x)$; let Σ_2 be the sphere in the space C_2 of radius $|W_0|_2$ around W_0, where $|\ |_2$ is the C_2 norm.[3] The transformation has these properties:

(a) It maps all sufficiently differentiable elements of Σ_2 into Σ_2, provided that the width of the strip in which we operate is less than δ_0. The choice of δ_0 depends only on the C_1 norm of A, B, T and W_0.

(b) τ is Lipschitz continuous in the <u>maximum norm</u> (C_0 norm) in Σ_2, i.e., if W and W' are a pair of elements in Σ_2 in the domain of τ,

$$|\tau(W) - \tau(W')|_0 \leq \lambda |W - W'|_0$$

λ can be made less than one by replacing the maximum norm by the <u>modified maximum norm</u>[4] (denoted by $|\ |_{mod}$):

$$|W|_{mod} = \max |e^{-kt} W(x,t)|_0$$

k sufficiently large.

To prove (a) we must get an estimate for the first and the second derivative of U in terms of $|W|_2$. To this end we differentiate (1.5) with respect to x and obtain a first order hyperbolic system in U_x; it can be diagonalized by introducing $Z = T^{-1}U_x$. Haar's lemma gives the desired estimate for Z and hence for U_x; the estimate for U_{xx} is derived from the twice differentiated equation.

To prove (b), one has to estimate the maximum difference of $U = \tau(W)$ and $U' = \tau(W')$; to accomplish this, subtract from each other (1.5) and (1.5'), the respective equations satisfied by U and U'. The result is a first order system in $U - U'$; diagonalize it and apply

---

[3] The C_k space is the space of k times differentiable vectors; the C_k norm of a vector is equal to the sum of the maxima of its first k derivatives.

[4] See K. O. Friedrichs, [9].

Haar's lemma. The result is the estimate stated in (b), and even more:

> τ is Lipschitz continuous in the maximum norm on any set bounded in the C_1 norm.

So far the transformation τ was defined only for sufficiently differentiable elements of the sphere Σ_2. But, since it is continuous in the C_0 norm, it can be defined at all points of the closure of Σ_2 in the sense of the C_0 norm. The closure of Σ_2 in the sense of the C_0 norm is the sphere Σ_{1+1}, consisting of all vectors W whose C_{1+1} distance[5] from W_0 is less than $|W_0|_2$: $|W-W_0|_{1+1} \leq |W_0|_2$. Properties (a) and (b) are preserved under this extension, i.e., τ maps Σ_{1+1} into itself and contracts in the sense of the C_0 norm. So iterates of the transformation will converge to a unique fixed point in Σ_{1+1} which is, as may be seen easily, a solution of (1) with initial value ϕ.

So far the existence of a solution was demonstrated only for sufficiently differentiable A, B and ϕ; but our results imply that a solution exists even if A, B and ϕ are only in C_{1+1}. Functions A, B and ϕ in C_{1+1} can be approximated uniformly by a sequence of smooth functions A_k, B_k, ϕ_k, whose C_{1+1} norm is uniformly bounded. According to our estimates, the corresponding solutions exist in a common strip, will be uniformly bounded in the C_{1+1} norm and will converge uniformly. Their limit, as may easily be seen, will be a solution of the original initial value problem.

It is more difficult to prove the existence of a solution if A, B and ϕ have merely continuous first derivatives. This was accomplished by A. Douglis (see [5]) with the aid of this estimate:

> If $w(\epsilon)$ is a modulus of continuity for the first derivatives of A, B and ϕ, then the first derivative of the corresponding solution $U(x,t)$ has a modulus of continuity of the form $k \cdot w(\epsilon)$, where the constant k depends only on the C_1 norm of A, B and ϕ.

This estimate cannot be derived from the Haar inequality. P. Hartman and A. Wintner, in [14], have rediscovered this result.

It is easy to show that if an initial vector belongs to class C_{1+1} or class C_1, then not only will the corresponding solution exist

---

[5] The space C_{1+1} consists of all vectors with Lipschitz continuous first derivatives; the C_{1+1} norm is the sum of the C_1 norm and the Lipschitz constant for the first derivative.

and belong to this differentiability class for a sufficiently short time interval, but the solution can lose these properties upon reaching a certain value of t only if its first derivative becomes unbounded. The same is true for a large class of higher differentiability properties.

Denote by $U = \mathcal{S}\phi$ the operator that assigns to each initial vector ϕ the corresponding solution U. $\mathcal{S}$ is defined for all ϕ in C_1 and has these properties:

If $\{\phi\}$ is a bounded set in C_1, there is a common strip in which the corresponding solutions $\mathcal{S}\phi$ exist and are uniformly bounded in the C_1 norm. Furthermore $\mathcal{S}\phi$ is continuous[6] in the C_0 norm in this strip over the set $\{\phi\}$.

These properties of $\mathcal{S}$ enable us to extend $\mathcal{S}$ to all Lipschitz continuous ϕ by closure in the C_0 norm. The image of a Lipschitz continuous ϕ under this extended transformation is called a generalized solution. The generalized solutions are Lipschitz continuous, and they can be continued in time until they cease to be Lipschitz continuous.

A Lipschitz continuous ϕ has a first derivative ϕ' almost everywhere. I was able to show, see [18]. that if ϕ' is continuous almost everywhere, then its discontinuities are propagated only along the characteristics, i.e., the first derivatives of $\mathcal{S}\phi$, which exist almost everywhere, are continuous except possibly at points which can be connected by a characteristic to a point of discontinuity of ϕ' on the initial interval.

The interest in generalized solutions and their properties is due partly to their occurrence in physical problems (initial data with discontinuous first derivatives) and partly because they throw additional light on the manner of the dependence of the solution on the initial data. It should be emphasized that we cannot expect to create by closure a sufficiently large class of generalized solutions on which a theory in the large could be built. In fact, as pointed out in the next section, we cannot, in general, extend by closure the domain of $\mathcal{S}\phi$ beyond the class of Lipschitz continuous ϕ, except for an exceptional class of systems.

§2. THEORY OF GENERALIZED SOLUTIONS (LIMITS OF GENUINE SOLUTIONS)

Further extension of $\mathcal{S}$ by closure is, in general, impossible for this reason: If an initial vector $\phi(x)$ is not Lipschitz continuous and is approximated by a sequence ϕ_i, then the first derivatives of ϕ_i cannot remain bounded, and hence the width of the strip in which the ith initial value problem has a genuine solution tends, in general, to zero

---

[6] See the proof of the corresponding result for the transformation τ , p. 173.

as i increases. Take as example the single nonlinear equation $u_t + uu_x = 0$. The solution of this nonlinear equation with a differentiable initial function $\phi(x)$ is given by the implicit relation $u = \phi(x-ut)$. This relation defines u as differentiable function of x and t only as long as the derivative $1 + t\phi'$ of $u - \phi(x-ut)$ with respect to u remains different from zero, i.e., if $t\phi'$ remains greater than minus one. Take now the discontinuous initial function

$$\phi(x) = \begin{matrix} a \text{ for } x < 0 \\ b \text{ for } x > 0 \end{matrix}$$

clearly if a is greater than b and ϕ_i is a sequence approximating ϕ, the minimum of ϕ_i' will tend to $-\infty$, and so the range of t for which the ith problem has a solution tends to zero. On the other hand, if a is less than b, the approximating sequence ϕ_i can be so chosen that u_i, the solution with initial value ϕ_i, will exist for all values of t, and, as is easy to see, the sequence $\{u_i\}$ converges to a continuous limit solution u which is differentiable for $t > 0$.

This example is probably typical of the general situation; there is, namely, strong indication that any vector which is the limit, in whatever topology, of genuine solutions of a given quasilinear system is <u>Lipschitz continuous</u> at every interior point of its domain of definition -- <u>unless the system belongs to a certain exceptional class</u>. This means that, in general, only those initial value problems can be solved by approximation through genuine solutions which have solutions Liptschitz continuous for positive t. Of course, there are solutions Lipschitz continuous for positive t, but not for $t = 0$ (for instance the above examples) just as there are Lipschitz continuous solutions which cannot be continued beyond a certain value of t. In general, the existence of such solutions is the <u>dual</u> of the phenomenon of the impossibility of continuing certain solutions beyond a critical value of t. More precisely, if $U_0(x,t)$ is a genuine solution of the quasilinear equation $U_t + AU_x + B = 0$, which ceases to be differentiable as t approaches t_0, then $U_0(x,t_0-t)$ is a solution of the <u>adjoint</u> equation $U_t - AU_x - B = 0$, Lipschitz continuous for $t > 0$ but whose initial value $\phi(x) = U_0(x,t_0)$ is not Lipschitz continuous. On the other hand, $\phi(x)$ is <u>not</u> the initial vector of any solution of the <u>original</u> equation which would be Lipschitz continuous for $t > 0$. This indicates that it is impossible to continue solutions beyond the breakdown point by approximating by genuine solutions.

In the next paragraphs we shall characterize the exceptional systems and present a few results on the manner of dependence of their solutions on initial data, and in particular results on discontinuous solutions of such systems.

The exceptional class is most easily characterized by means of a <u>normal form</u> for quasilinear hyperbolic systems which can be achieved by introducing new independent variables:

Multiply the equation $U_t + AU_x + B = 0$ on the left by E_i, the ith left eigenvector of the matrix A, $E_i A = d_i E_i$. The resulting equations can be written as

$$E_i \frac{dU}{di} + b_i = 0, \quad i = 1,2,\ldots,n \tag{2.1}$$

where

$$\frac{dU}{di} = U_t + d_i U_x$$

denotes the directional derivative of U in the ith characteristic direction.

In equations (2.1) each unknown appears differentiated in a single direction. In the linear and semilinear case, i.e., when E_i is independent of U, it is possible to introduce new unknowns, $v^i = E_i U$ such that the ith equation in (2.1) will involve the derivative of v^i only. The introduction of such characteristic variables is still possible for quasilinear systems in <u>two</u> unknowns. Write out the equations (2.1) longhand:

$$e_{i1} \frac{du^1}{di} + e_{i2} \frac{du^2}{di} + b_i = 0, \quad i = 1,2\ . \tag{2.2}$$

There exist integrating factors f_i, $i = 1,2$, i.e., factors such that $f_i e_{i1}$ and $f_i e_{i2}$ are, respectively, the u^1 and u^2 derivatives of the new functions v^i of x,t,u^1,u^2. Clearly, after multiplying (2.2) by f_i, it can be written as

$$\frac{dv^i}{di} + c_i = 0, \quad i = 1,2 \tag{2.3}$$ [7]

For n greater than two no integrating factor exists in general, and this precludes the existence of characteristic variables. This much, however, may be accomplished:

Given any $n-1$ of the n equations (2.1) - say the first $n-1$ - it is possible to introduce new unknowns so that the first $n-1$ equations in (2.1) contain derivatives of $n-1$ unknowns only. This is how the new unknowns are chosen: denote by $L = (l_1,\ldots,l_n)$ a vector orthogonal to

[7] These functions v^1, v^2 are called generalized Riemann invariants, see [10].

$E_1, E_2, \ldots, E_{n-1}$. (Because of the biorthogonality of the eigenvectors of a matrix and its adjoint, L is the right eigenvector of A with eigenvalue d_n.) L is of course a function of x, t, U. For any fixed value of these variables, the vectors $E_1, \ldots, E_{n-1}$ can be expressed as linear combinations of any $n-1$ linearly independent vectors F_i, $i = 1, 2, \ldots, n-1$, orthogonal to L. We shall choose the F_i to depend in a special way on the variables, namely to be gradients with respect to $u^1, u^2, \ldots, u^n$: $F_i = \operatorname{grad} v^i$, $i = 1, 2, \ldots, n-1$. The condition that F_i be orthogonal to L,

$$\sum_{j=1}^{n} l_j v^i_{u^j} = 0 \tag{2.4}$$

is a linear first order homogeneous equation for the functions v^i. According to the theory of such equations - see e.g., [3], pp. 23-25 - all solutions can be expressed as arbitrary functions of $k-1$ independent solutions, k being the number of independent variables. In our case the independent variables are x, t and U, so k is $n+2$. x and t are two solutions of (2.4), independent of each other, so $n-1$ further independent solutions may be found. Let $v^1, \ldots, v^{n-1}$ be any such independent solutions.

It is easy to see that the directional derivatives of U occurring in the first $n-1$ equation of (2.1) can be expressed as linear combination of directional derivatives of $v^1, \ldots, v^{n-1}$.

> DEFINITION: A system is called exceptional if d_n, the slope of the nth characteristic, can be expressed as a function of $x, t, v^1, v^2, \ldots, v^{n-1}$.

Analytically, this means that the gradient of d_n with respect to U should be expressible as a linear combination of the gradients of $v^1, v^2, \ldots, v^{n-1}$ with respect to U. This is the same as saying that grad d_n is orthogonal to L, which leads to this useful criterion:

> The system $U_t + AU_x + B = 0$ is exceptional with respect to the nth characteristic field if and only if the gradient of d_n (the nth eigenvalue of A) is orthogonal to L_n (the right eigenvector of A with eigenvalue d_n).

The eigenvalue d_n is supposed to be simple, and the orthogonality is supposed to hold for all relevant values of x, t and U.

Imagine new unknowns $v^1, v^2, \ldots, v^{n-1}$ introduced according to the recipe given before, and an nth one, v^n, any old way, as long as it is independent of $x,t,v^1,v^2,\ldots,v^{n-1}$. We introduce two norms for initial vectors Φ :

$$|\,\Phi\,| = \sum_{i=1}^{n-1} |\phi^i|_l + |\phi^n|_b$$

$$|\,\Phi\,|' = \sum_{i=1}^{n-1} |\phi^i|_0 + |\phi^n|_L$$

We shall refer to the norms as the "unprimed" and "primed" norm. $|\phi^i|_l$ denotes here the Lipschitz norm, $|\phi^n|_b$ the total variation of ϕ^n over its interval of definition, $|\phi^i|_o$ the maximum norm and $|\phi^n|_L$ the L_1 norm.

The next three theorems state some of the properties of exceptional systems:

THEOREM I: The width of the strip in which the solution of an initial value problem $V(x,0) = \Phi$ exists[8] is at least as large as $\mathrm{const}\,|\Phi|^{-1}$; the value of the constant depends on the magnitude of the coefficients A,B,T and their first derivatives.

Denote by $S(t_0)$ the transformation that links the initial vector Φ of a solution $V(x,t)$ to its value Ψ on the line $t = t_0: \Psi = S(t_0)\,\Phi$.

THEOREM II: $S(t)$ is a bounded transformation in the sense of the unprimed norm.

THEOREM III: In any set of vectors $\{\Phi\}$ bounded in the unprimed norm, S is a continuous transformation in the sense of the primed norm.

These theorems enable us to construct by closure with respect to the primed norm solutions for initial vectors whose first n-1 components are Lipschitz continuous, but whose last component need be only of bounded

[8] In [22], Hans Lewy investigates the magnitude of the domain of existence of an initial value problem in the plane of the characteristic parameters for systems where there are only two characteristic directions.

variation; in particular the last component may be discontinuous. Proofs of these theorems will be published elsewhere.

The equations of one-dimensional, time-dependent, non-isentropic, compressible flow are examples of exceptional systems; these results presented above can be used to show that flows of the above kind with contact discontinuities can be obtained as limits of flows without discontinuities.

§3. THEORY OF WEAK SOLUTIONS

The theory in the large is based on the concept of weak solution, which may be defined only for a special class of quasilinear hyperbolic systems, for those namely which are written in divergence form (equations in this form are called conservation laws):

$$U_t + F_x(x,t,U) + B = 0 \tag{3.1}$$

F is some nonlinear vector function of x,t,U. U is called a weak solution of (3.1) with initial value Φ if the integrated equation

$$- \int W(x,o)\, \Phi\, dx + \iint \left\{ -W_t U - W_x F + WB \right\} dx\, dt = 0$$

obtained by multiplying (3.1) by W on the left, integrating the resulting equation, and performing two integrations by parts, holds for every test vector W which has continuous first derivatives and which vanishes outside of a bounded set.

The concept of weak solution is associated with the form in which the equations are written (as long as they are written in divergence form) and not with the equations themselves.[9] That is, suppose that new variables V can be introduced as certain nonlinear functions of x,t,U:

$$V = H(x,t,U) \tag{3.2}$$

so that when the system (3.1) is rewritten in terms of V and solved for V_t, the ensuing system is again in divergence form:

$$V_t + G_x(x,t,V) + C = 0 \tag{3.3}$$

Weak solutions of (3.3) are defined the same way as for (3.1). If $U_0(x,t)$ is a weak solution of (3.1), then $V_0 = H(x,t,U_0)$ will not be

[9] This indicates that only those weak solutions which do not depend on the form of the equations are obtainable as limits of genuine solutions. Notice also that the normal form (2.1) loses its significance.

a weak solution of (3.3) unless U_0 is Lipschitz continuous or the system is of the exceptional class. The equations of one dimensional fluid dynamics in Lagrange mass variable furnish an example of the phenomenon. The four equations expressing the time derivative of specific volume, velocity, total energy per unit mass (i.e., kinetic plus internal) and entropy per unit mass are all in divergence form. But, whereas a genuine solution of the first three is also a genuine solution of the fourth, a weak solution of the first three would not, in general, be a weak solution of the fourth.

The main questions are:

(a) What is the class of vectors that may serve as initial values of weak solutions?

(b) Are two weak solutions with the same initial values identical?

The answer to (a) is not known in general. E. Hopf has shown (see [16]) that the equation

$$u_t + (u^2/2)_x = 0$$

has a weak solution with arbitrarily prescribed bounded measurable initial function. Presumably something similar is true for other equations.

The answer to (b) is an emphatic no! For example, the equation

$$u_t + (u^2/2)_x = 0$$

has infinitely many weak solutions with the initial value

$$\phi(x) = \begin{cases} 0 & \text{for } x < 0 \\ 1 & \text{for } x > 0 \end{cases}$$

The functions

$$u(x,t) = \begin{cases} 0 & \text{for } x \leq 0 \\ x/t & \text{for } 0 < x \leq t \\ 1 & \text{for } 0 < t \leq x \end{cases}$$

and

$$u(x,t) = \begin{cases} 0 & \text{for } 2x < t \\ 1 & \text{for } 2x > t \end{cases}$$

are two examples, and there are infinitely many others. Nevertheless, if we accept the idea that the solution of the initial value problem describes the progress of a physical system, there must be some additional principle (or various additional principles) which selects a certain weak solution of each initial value problem as "physically relevant".

Such a guiding principle is Hadamard's postulate that in a

physical problem the solution depends continuously on the data; in our case this means that small changes in the initial vector should result in small changes of the corresponding physically relevant solution. My conjecture is that this requirement characterizes the class of physically relevant solutions uniquely. More precisely:

> CONJECTURE: Among all functions $U = R(\Phi)$ which assign to each vector Φ a weak solution U with initial value Φ, there exists exactly one which is continuous in a suitable topology.

In case of the equation $u_t + (u^2/2)_x = 0$, it can be shown that there exists <u>at least one</u> such assignment $u = R(\phi)$, namely the one constructed by the viscosity method: Enlarge the equation by the additional term λu_{xx} on the right, obtaining[10]

(3.4) $$u_t + (u^2/2)_x = \lambda u_{xx}, \quad u(x,0) = \phi(x)$$

E. Hopf has shown (see [13]) that if ϕ is any bounded measurable function, this nonlinear parabolic equation has a solution u_λ which exists for all $t > 0$; as λ tends to zero, $u_\lambda(x,t)$ converges for almost all x and t to a limit function $u(x,t)$. Since the u_λ are uniformly bounded, $u_\lambda \longrightarrow u$, $u_\lambda^2 \longrightarrow u^2$ in the L_1 sense in every finite subdomain; consequently, since u_λ is at the same time a <u>weak</u> solution of (3.4) with initial value ϕ, it follows that u is a weak solution of the original equation with initial value ϕ.

It is not hard to show, pursuant to Hopf's work, that the weak solution u assigned to each bounded measurable ϕ depends continuously on ϕ in the weak topology; more precisely, if $\phi_i(x)$ is a uniformly bounded sequence for which

$$\int_0^x \phi_i dx$$

tend to

$$\int_0^x \phi \, dx$$

then u_i tends to u in the same sense for every fixed t.

The viscosity method can be applied to the general situation (3.1): By inserting the term λU_{xx} on the right one obtains an initial value problem for a nonlinear parabolic system

[10] Equation (3.4) was first discussed by J. M. Burgers.

(3.4') $$U_x + F_x + B = \lambda U_{xx}, \quad U(x,0) = \Phi(x)$$

So far the relevant theorems concerning the existence of these solutions in the large and their convergence as λ tends to zero are missing. It would follow however by the same reasoning as in the special case of equation (3.4) that a strong limit of solutions of (3.4') is a weak solution of (3.4).

The convergence of steady state solutions of (3.4') to steady state weak solutions of (3.1) has been investigated by many authors, (see e.g., [2], section 63). In these investigations Φ is taken to be constant for negative x and another constant for positive x. The two constant states are so chosen that there exists a weak solution U with initial value Φ which just translates Φ at constant speed: $U(x,t) = \Phi(x-ct)$. These investigations show the existence of a sequence of vectors Φ_λ, approaching Φ as λ tends to zero, which are translated at the same constant speed by the solution U_λ of (3.4'):

$$U_\lambda(x,t) = \Phi_\lambda(x-ct)$$

so that U_λ tends to U.

It should be pointed out that the class of weak solutions constructed by the viscosity method is <u>irreversible in time</u>. That is if the value of the weak solution of $U_t + F_x + B = 0$ with initial value Φ constructed by the viscosity method, is equal to Ψ at $t = t_0$, the value of the weak solution of $U_t - F_x - B = 0$ with initial value Ψ constructed by the viscosity method may very well <u>not</u> be equal to Φ at $t = t_0$. There is nothing surprising in this, for the parabolic equations employed in the limiting process distinguishes plus t from minus t.

The following result sheds some light on the degree of irreversibility. Denote by $u = R(\phi)$ the weak solution of $(u^2/2)_x = 0$ assigned by the viscosity method to each bounded measurable ϕ.

> THEOREM: $R(\phi)$ is completely continuous, i.e., if ϕ_i is a uniformly bounded sequence of initial functions, then the corresponding $u_i = R(\phi_i)$ uniformly bounded and contain a subsequence which converges in the L_1 sense in every bounded subdomain of the upper halfplane.

Proof of this theorem will be published separately; it rests on the continuity of $R(\phi)$ in the weak topology, and this principle: If a <u>weak limit</u> of a sequence of weak solutions u_i of a nonlinear (specifically quadratic) equation is itself a weak solution, then it must be a strong limit of the sequence $\{u_i\}$.

Let us turn now to the second part of the conjecture, that there is at most one continuous $S(\Phi)$. There is no proof of this, not even in the simplest cases. I am proposing it merely as a possible explanation of what I observed while investigating other limiting procedures -- different from the method of viscosity -- which would assign to each Φ a weak solution with initial value Φ. The method used[11] was the following finite difference scheme:

In the differential equation (3.1) replace the space derivative $F_x(x,t,U(x,t))$ by the symmetric difference quotient

$$(F(x+\Delta x,t,U(x+\Delta x,t)) - F(x-\Delta x,t,U(x-\Delta x,t)))/2\Delta x,$$

the time derivative U_t by the forward difference quotient

$$(2U(x,t+\Delta t) - U(x+\Delta x,t) - U(x-\Delta x,t))/2\Delta t$$

The resulting difference equation enables us to calculate $U(x,t+\Delta t)$ if $U(x,t)$ is known. In particular, if $U(x,0)$ is known, we can determine $U(x,t)$ for all values of t which are integer multiples of Δt (the value of $U(x,0)$ at the lattice points $x = m\Delta x$, $m = 0, \pm 2, \ldots$ determines the value of $U(x,t)$ at all points of a staggered lattice).

Denote by Δ a particular meshwidth $\Delta = (\Delta t, \Delta x)$, and by U_Δ the corresponding solution of the finite difference equations with initial value Φ. $U_\Delta(x,t)$ is defined only if t is an integer multiple of Δt, but we define it for all other values by taking $U_\Delta(x,t)$ equal to $U_\Delta(x, \nu\Delta t)$ for $\nu\Delta t < t < (\nu+1)\Delta t$, ν integer. Suppose that as Δt and Δx tend to zero in some prescribed fashion, the corresponding solutions converge in, say, the L_1 sense to a limit U. It is easy to show that this limit U is a weak solution.[12]

Experimental calculations tried on various discontinuous initial values for the equation $u_t + (u^2/2)_x = 0$, and the equations of one-dimensional hydrodynamics, both Euler and Lagrange forms (using, in the first case, mass, momentum and total energy per unit volume as unknowns; in the second case volume, momentum and total energy per unit mass), definitely indicate that if Δx and Δt tend to zero so that $\Delta t/\Delta x$ does not exceed the absolute value of the slope of all characteristics (which

---

[11] The numerical results obtained by using this method are described in [19]; there is much in common between the method described here and the ones proposed by J. v. Neumann in [24] and by J. v. Neumann and R. D. Richtmyer in [25].

[12] If the convergence is in the weak sense only, the weak limit would in general not be a weak solution.

is the classical Courant-Friedrichs-Lewy criterion) then the sequence U_Δ does indeed converge.

There is no mathematical proof for this convergence except in the small, for the conventional case of differentiable Φ ; but -- if the result is true -- it should be possible, at the present advanced state of analysis, to give a rigorous proof of it. Nor can it be much harder to prove -- in view of the systematic nature of the method -- that the weak solution U assigned to each Φ by this method depends continuously on Φ. The experimental evidence bears this out; even more: the weak solutions obtained by this method seem to be the same as the ones obtained by the viscosity method. A plausible explanation for this phenomenon seems to be this -- based on the conjecture stated before --: Since both methods assign a weak solution U to each vector Φ so that U depends continuously on Φ, they must assign the same solutions.

REMARK: It is to be expected that the class of weak solutions furnished by the finite difference scheme described here is irreversible in time since the scheme employs forward differences in time and so distinguishes plus t from minus t.

It would be interesting to investigate other types of limiting procedures, and to see whether they produce the same class of weak solutions as the previous methods. The proper abstract setting for these investigations may be the class of equations

$$U_t = ANU + MU$$

where A is an unbounded linear, N and M are continuous nonlinear transformations over a Banach space.

BIBLIOGRAPHY

[1] CINQUINI-CIBRARIO, M., "Un teorema di esistenza e di unicità per un sistema di equazioni alle derivate parziale," Annali di Matematica 24 (1945), 157-175.

[2] COURANT, R., FRIEDRICHS, K. O., Supersonic Flow and Shock Waves, Interscience, New York (1948).

[3] COURANT, R., HILBERT, D., Methoden der Matematischen Physik 2, Springer, Berlin (1937).

[4] COURANT, R., LAX, P. D., "Nonlinear partial differential equations with two dependent variables," Communications on Pure and Applied Mathematics 2 (1949), 255-273.

[5] DOUGLIS, A., "Existence theorems for hyperbolic systems," Communications on Pure and Applied Mathematics 5 (1952), 119-154.

BIBLIOGRAPHY

[6] FRIEDRICHS, K. O., LEWY, H., "Das Anfangswertproblem einer beliebigen nichtlinearen hyperbolischen Differenzialgleichung beliebiger Ordnung in zwei Variablen. Existenz, Eindeutigkeit und Abhängigkeitsbereich der Lösung," Mathematische Annalen 99 (1928), 200-221.

[7] FRIEDRICHS, K. O., LEWY, H., "Ueber fortsetzbare Anfangsbedingungen bei hyperbolischen Differenzialgleichungen in drei Veränderlichen," Nachrichten der Gesellschaft der Wissenschaft zu Göttingen, Mathematische Physikalische Klasse No. 26 (1932), 135-143.

[8] FRIEDRICHS, K. O., "On hyperbolic differential operators," Bulletin of the American Mathematical Society, abstract 46-9-416 (1940), 754-

[9] FRIEDRICHS, K. O., "Nonlinear hyperbolic differential equations for functions of two independent variables," American Journal of Mathematics 70 (1948), 555-589.

[10] HAACK, W., HELLWIG, G., "Ueber Systeme hyperbolischer Differenzialgleichungen erster Ordnung I," Mathematische Zeitschrift 53 (1950), 244-266.

[11] HAAR, A., "Sur l'unicité des solutions des équations aux dérivées partielles," Comptes Rendus des Séances de l'Académie des Sciences 187 (1928), 23-

[12] HAAR, A., "Ueber Eindeutigkeit und Analyzität der Lösungen partieller Differentialgleichungen," Atti del Congresso Internazionale dei Matematici, Bologna (1928), Volume III, 5-10.

[13] HADAMARD, J., "Observations sur la note précédante," Comptes Rendus des Séances de l'Académie des Sciences 187 (1928), 23-

[14] HARTMAN, P., WINTNER, A., "On hyperbolic partial differential equations," American Journal of Mathematics 74 (1952), 834-864.

[15] HOLMGREN, E., "Sur la théorie des équations intégrales linéaires," Arkiv för Matematik, Astronomi och Fysik 3 (1906).

[16] HOPF, E., "The partial differential equation $u_t + uu_x = \mu u_{xx}$," Communications on Pure and Applied Mathematics 3 (1950), 201-230.

[17] LAX, P. D., "Completely continuous transformations," Bulletin, American Mathematical Society 57 (1951), 274-

[18] LAX, P. D., "Nonlinear hyperbolic equations," Communications on Pure and Applied Mathematics 6 (1953),

[19] LAX, P. D., "On discontinuous initial value problems for nonlinear equations and finite difference schemes," Los Alamos Report LAMS 1332 (1952).

[20] LERAY, J., "Hyperbolic equations with variable coefficients," Lectures delivered at the Institute for Advanced Study, Princeton (1952).

[21] LEWY, H., "Ueber des Anfangswertproblem bei einer hyperbolischen nichtlinearen partiellen Differenzialgleichung zweiter Ordnung mit zwei unabhängingen Veränderlichen," Mathematische Annalen 98 (1927), 179-191.

[22] LEWY, H., "Generalized integrals and differential equations," Transactions, American Mathematical Society 43 (1938), 437-464.

[23] LEWY, H., "A priori limitations for solutions of Monge-Ampère equations I," Transactions, American Mathematical Society 37 (1935), 417-434.

BIBLIOGRAPHY

[24] v. NEUMANN, J., "Proposal and analysis of a numerical method for the treatment of hydrodynamic shock problems," NDRC Report AM 551 (1944).

[25] v. NEUMANN, J., RICHTMYER, R. D., "A method for numerical calculation of hydrodynamic shocks," Journal of Applied Physics 21 (1950, 232-237.

[26] PERRON, O., "Ueber Existenz und Nichtexistenz von Integralen partieller Differenzialgleichungssysteme im reellen Gebiet," Mathematische Zeitschrift 27 (1928), 549-564.

[27] PETROWSKY, I., "Ueber das Cauchysche Problem für Systeme von partiellen Differenzialgleichungen," Recueil Mathématique 2 (1937), 815-858.

[28] SCHAUDER, J., "Cauchysches Problem für partielle Differenzialgleichungen erster Ordnung. Anwendung einiger sich auf die Absolutbeträge der Lösungen beziehenden Abschätzungen," Commentarii Mathematici Helvetici 9 (1937), 263-283.

XIII. A GEOMETRIC TREATMENT OF LINEAR HYPERBOLIC EQUATIONS OF SECOND ORDER

A. Douglis

In this paper, we shall describe an elementary theory of linear, hyperbolic equations of the second order with variable coefficients which is applicable to a variety of questions concerning initial-value problems with or without boundary conditions. The solutions of these problems are characterized as the solutions of certain integral equations of Volterra type which, expressed in suitable coordinates, are of relatively simple structure and thus lend themselves to the detailed study of their solutions.[1]

One outcome of the new approach is a proof[2] of a conjecture of Hadamard's that the linear, hyperbolic equations of second order in three plus one independent variables for which Huygens' Principle is valid are just those which result from the wave equation after transformations of three possible kinds: change of independent variables, multiplication of the dependent variable by a variable factor, and multiplication of each term of the equation by a variable factor.[3]

In all the problems mentioned, we begin by reducing an equation of the type considered to a certain normal form. Corresponding to any point $Q: x^i = \xi^i, \quad i = 0,1,\ldots,n,$ there are transformations of these three kinds leaving Q invariant, under which the equation takes the normal form

$$Mu \equiv \frac{\partial}{\partial y^i}\left(h^{ij}(y;Q)\,\frac{\partial u}{\partial y^j}\right) + k^i(y;Q)\,\frac{\partial u}{\partial y^i} + f(y;Q)u = g(y;Q) \quad {}^4$$

---

[1] The integral equations that arise in Cauchy's problem are of the type given by M. Mathisson [6], S. Sobolev [7], [8], and by Y. Fourès-Bruhat [3]; they are here somewhat differently derived and expressed.

[2] To be published in the Communications on Pure and Applied Mathematics.

[3] A special case of the theorem has been given by Mathisson [5] and by Hadamard [4].

[4] Summation from 0 to n over repeated indices.

distinguished by the following properties:

(1) The line $y^1 = \xi^1, \ldots, y^n = \xi^n$ is a bicharacteristic line; on it, the leading coefficients satisfy the conditions $h^{00} = 1$, $h^{11} = h^{22} = \ldots = h^{nn} = -1$, $h^{ij} = 0$ $(i \neq j)$,

$$\frac{\partial h^{kl}}{\partial y^m} = 0 \qquad (k,l,m = 0, \ldots, n)$$

(2) Identically, $\det(h^{ij}(y;Q)) = -1$.

(3) The cones $\theta(y) \equiv y^0 + r = \text{const.}$,

$$r = \sqrt{(y^1-\xi^1)^2 + \ldots + (y^n-\xi^n)^2}$$

are characteristic surfaces, and the generators of the cones are bicharacteristic lines.

Our use of the normal form is based on a geometrical principle which is a sharpened statement of the well-known property that the second-order terms of a hyperbolic equation of second order constitute an "inner" differential expression with respect to a characteristic surface. This principle takes the form of an identity, which generalizes a formula given by H. Lewy[5] for the wave equation in four dimensions.[6]

Let $I : y^0 = T(y^1, \ldots, y^n)$ be a space-like initial manifold. Let

$$C^\tau : \quad \theta(y) \equiv y^0 + r = \tau, \qquad r = \sqrt{(y^1-\xi^1)^2 + \ldots + (y^n-\xi^n)^2}$$

be the characteristic cone with vertex

$$y^0 = \tau, \quad y^1 = \xi^1, \ldots, y^n = \xi^n$$

let K^τ be the cylindrical image of this cone under the transformation

$$Y^0 = y^0, \quad Y^1 = \frac{y^1-\xi^1}{r}, \ldots, \quad Y^n = \frac{y^n-\xi^n}{r}$$

K^τ_ϵ $(\epsilon > 0)$ the truncated piece of K^τ for which

---

5 cf. Courant-Hilbert [2], pp. 370-371.

6 The principle is implicitly contained in a work of Beltrami's [1] published in 1892, dealing with the wave equation in four dimensions. It is inherent also in the treatments of Cauchy's problem for hyperbolic equations with variable coefficients given by Mathisson [6] and by Sobolev [8], whose methods are also explained by Fourès-Bruhat [3] in connection with her treatment of non-linear equations. The most penetrating formulation of this principle, presented in a special case in the shape of an identity, is due, however, to Lewy as indicated in the text.

$$T(y^1,\ldots,y^n) \leq y^0 \leq \tau - \epsilon$$

and K_0^τ the part of K^τ for which

$$T(y^1,\ldots,y^n) \leq y^0 < \tau$$

We shall mean by the surface integral of a function W over K_0^τ the limit (if it exists) as $\epsilon \longrightarrow 0$ of the surface integrals

$$\int_{K_\epsilon^\tau} W \, dS \qquad \left(dS = \frac{dY^1 \ldots dY^{n-1}}{\sqrt{1 - (Y^1)^2 - \ldots - (Y^{n-1})^2}} \cdot dY^0 \right)$$

of W over K_ϵ^τ. Further, let us say that a given expression is of class I, if its surface integral over K_0^τ exists and is determined solely by data on I; such an expression we designate generically as (I). In this terminology, the fundamental identity, easily derived after changing to new coordinates

$$z^0 = -(y^0 - \xi^0), \quad z^1 = \frac{y^1 - \xi^1}{r}, \quad \ldots, \quad z^{n-1} = \frac{y^{n-1} - \xi^{n-1}}{r}, \quad z^n = \theta(y)$$

can now be stated as follows: if u, v are functions of (y^i) of class C^2, and if $v(\tau, \xi^1, \ldots, \xi^n) = 0$, then

$$vMu = \frac{\partial}{\partial z^0}(Au) + B\frac{\partial u}{\partial z^n} + Cu + (I)$$

where

$$A = -\frac{\partial v}{\partial z^n}, \quad B = 2\frac{\partial v}{\partial z^0} - \left(\frac{n-1}{r} - \alpha_i k^i\right) v, \quad \alpha_0 = 1$$

$$\alpha_1 = z^1, \quad \ldots, \quad \alpha_{n-1} = z^{n-1}, \quad \alpha_n = \sqrt{1 - \alpha_1^2 - \ldots - \alpha_n^2}$$

and C is a function continuous on K_0^τ.

Cauchy's problem in four dimensions $(n = 3)$, for example, is now attacked with the aid of this identity by setting

$$v = r \exp \left\{ -\frac{1}{2}\alpha_i \int_o^r k^i(\tau - \rho, \xi^1 + \alpha_1\rho, \ldots, \xi^n + \alpha_n\rho)\, d\rho \right\}$$

(for which choice $A \neq 0$, $B = 0$) and forming the surface integrals over $K_0^{\xi^0}$ of both sides of the resulting formula. An integral relation of

Volterra type is obtained which reduces, in the case of the wave equation, to Poisson's well-known representation. Cauchy's problem for any value of n and other types of problems mentioned can be solved by extending this procedure, as will be fully discussed in forthcoming papers.

BIBLIOGRAPHY

[1] BELTRAMI, E., "Sull'espressione analitica del principio di Huygens," Rendiconti della R. Accademia dei Lincei 1 (1892 first semester), 99-108. Republished in "Opere Matematiche di Eugenio Beltrami," 4 Ulrico Hoepli, Milan (1920), 499-510.

[2] COURANT, R., HILBERT, D., Methoden der Mathematischen Physik, 2, Ch. 6, Springer, Berlin (1937). Photo-lithoprinted by Interscience Publishers, New York.

[3] FOURÈS-BRUHAT, Y., "Théorème d'existence pour certains systèmes d'équations aux dérivées partielles non-linéaires," Acta Mathematica 88 (1952), 141-225.

[4] HADAMARD, J., "The problem of diffusion of waves," Annals of Mathematics 2 43 (1942), 510-522.

[5] MATHISSON, M., "Le problème de M. Hadamard relatif à la diffusion des ondes," Acta Mathematica 70 (1939), 249-282.

[6] ——————, "Eine neue Lösungsmethode für Differentialgleichungen von normalem hyperbolischem Typus," Mathematische Annalen 107 (1932), 400-419.

[7] SOBOLEV, S., "Sur une généralisation de la formule de Kirchoff," Akademiya Nauk Doklady, N.S., No. 6, (1933), 258-262.

[8] ——————, "Méthode nouvelle à résoudre le problème de Cauchy pour les équations linéaires hyperboliques normales," Matheaticheskii Sbornik N.S. 1 43, (1936), 39-71.

XIV. ON CAUCHY'S PROBLEM AND FUNDAMENTAL SOLUTIONS*

J. B. Diaz

INTRODUCTION

One of the chief aims of the Arden House Conference on Partial Differential Equations was to bring together workers in the field from various geographically scattered schools of mathematicians in this country and abroad for the personal communication and dissemination of recent advances achieved in their respective groups. The present brief report is written in accordance with the avowed purpose of the Conference understood as above.

During the past several years, a group of mathematicians at the Institute for Fluid Dynamics and Applied Mathematics at the University of Maryland has been interested in certain questions in the theory of partial differential equations which may be broadly classified under the title headings of the three following sections. The purpose of the present article is to review some of the latest results obtained by this group. In a summary account of this kind it is of course impossible to enter into a luxury of detail, and only the bare outline of the results will be presented. The references in the bibliography contain a fuller discussion of the topics mentioned below.

§1. RIEMANN'S METHOD AND GENERALIZATION OF THE LAGRANGE-GREEN IDENTITY

Riemann's method for solving Cauchy's problem for the equation

$$L(u) = u_{xy} - au_x - bu_y = 0, \quad a = a(x,y), \quad b = b(x,y)$$

is based essentially on the bilinear Lagrange differential identity

$$vL(u) - uM(v) = A_x + B_y \tag{1}$$

where

$$M(v) = v_{xy} + (av)_x + (bv)_y$$

* This work was sponsored by the Office of Naval Research.

is the Lagrange adjoint of the operator L, and A and B are certain bilinear forms in u, v, u_x, v_x, u_y, v_y. The identity (1) implies that the line integral

$$I = \int \{B\,dx - A\,dy\} \tag{2}$$

vanishes when extended over closed paths lying in the interior of a domain D on which the regular functions u and v satisfy $L(u) = 0$ and $M(v) = 0$. Consider the Cauchy problem for $L(u) = 0$, Cauchy data being given on a curve C. Let v be a certain two parameter family of solutions of the adjoint equation $M(v) = 0$. The fact that I vanishes when taken around a curvilinear triangle $PXYP$ (here P is a point sufficiently close to C, but not on C, and X and Y are points on C) formed by the curve C and the characteristics PX and YP of the operator L leads to the following formula for the value of the solution u of the Cauchy problem at the point P:

$$u(P) = \frac{1}{2}\,[u(X)v(X) + u(Y)v(Y)] + \int_X^Y \{B\,dx - A\,dy\} \tag{3}$$

the line integral being taken along the segment of C between the points X and Y. Thus the solution of Cauchy's problem is reduced to the problem of the determination of the Riemann function v.

M. H. Martin [15] has recently given an alternative approach to the solution of Cauchy's problem for $L(u) = 0$. Martin's starting point is (instead of Lagrange's identity (1)) the following bilinear divergence identity:

$$\left(\frac{v_x}{\phi_x} - \frac{v_y}{\phi_y}\right) L(u) + \left(\frac{u_x}{\phi_x} - \frac{u_y}{\phi_y}\right) M(v) = \left(\frac{u_x v_x}{\phi_x}\right)_y - \left(\frac{u_y v_y}{\phi_y}\right)_x \tag{4}$$

where ϕ is a solution of $L(u) = 0$ for which $\phi_x \neq 0$, $\phi_y \neq 0$, and

$$M(v) = v_{xy} - b\,\phi_x^{-1}\,\phi_y v_x - a\,\phi_x\,\phi_y^{-1}\,v_y$$

This operator M, called the <u>associate operator</u>, takes the place of the Lagrange adjoint M in (1). There is an identity (4) for each solution ϕ of $L(u) = 0$, $\phi_x \neq 0$, $\phi_y \neq 0$; and, automatically, $M(\phi) = 0$. From (4), the line integral

$$I = \int \left\{\phi_x^{-1} u_x v_x\,dx + \phi_y^{-1} u_y v_y\,dy\right\} \tag{5}$$

vanishes around closed paths lying in a region D where u and v are regular functions satisfying $L(u) = 0$ and $M(v) = 0$. To solve the Cauchy problem considered above, a certain two parameter family v of solutions of the associate equation $M(v) = 0$, termed the ϕ-resolvent (the analogue to Riemann's function above), is required; and the vanishing of I in (5) when taken around the curvilinear triangle PXYP described before leads to the solution

$$u(P) = \frac{1}{2}[u(X) + u(Y)] + \frac{1}{2}\int_X^Y \left\{\phi_x^{-1}u_x v_x \, dx + \phi_y^{-1}u_y v_y \, dy\right\} \tag{6}$$

of Cauchy's problem, in place of (3).

Thus the solution of Cauchy's problem is reduced to the determination of the "ϕ-resolvent" v. This whole complex of ideas stems from an earlier paper of M. H. Martin [14] where a particular equation

$$L(u) = u_{xy} - \frac{m}{x - y}(u_x - u_y) = 0$$

of Euler-Poisson type was considered. A further extension of this method to cover the case of elliptic and parabolic equations in two independent variables, not necessarily in normal form, has been carried out by Diaz and Martin [7].

The extension of these ideas to more than two independent variables, i.e. the use of bilinear divergence identitites of type (4) instead of the usual Lagrange identity (1), seems to offer interesting possibilities. J. B. Diaz and M. H. Martin [6] obtained in this manner the known solution of the Cauchy problem for the wave equation

$$u_{x_1x_1} + \dots + u_{x_nx_n} - u_{tt} = 0 \tag{7}$$

Cauchy data being assigned on the plane $t = 0$. In this process it is not necessary to distinguish between the cases of n even and n odd in the course of the argument. The Cauchy problem for equation (7) with $n = 2$ has been treated earlier by Martin [15]. This method is different from an earlier extension of Riemann's method given by H. Lewy [13], who uses three "Riemann functions" for $n = 2$.

In order to illustrate the method, take $n = 3$ in (7), which simplifies the writing. The Cauchy problem is then

$$\begin{aligned} &L(u) = u_{xx} + u_{yy} + u_{zz} - u_{tt} = 0, \quad t > 0 \\ &u(x, y, z, 0) = f(x, y, z), \quad u_t(x, y, z, 0) = g(x, y, z) \end{aligned} \tag{8}$$

where f and g are given functions. The problem consists in the determination of the value of the function u at any point $(\overline{x}, \overline{y}, \overline{z}, \overline{t})$, where $\overline{t} > 0$, in terms of the assigned Cauchy data.

Let r, ρ, θ, t where ρ is the colatitude and θ is the longitude, be cylindrical coordinates in x, y, z, t space, that is

$$x - \overline{x} = r \cos \rho, \qquad r \geqq 0$$

$$y - \overline{y} = r \sin \rho \cos \theta, \qquad 0 \leqq \rho \leqq \pi$$

$$z - \overline{z} = r \sin \rho \sin \theta, \qquad 0 \leqq \theta < 2\pi$$

In these coordinates

$$(9) \qquad L(u) = u_{xx} + u_{yy} + u_{zz} - u_{tt}$$

$$= u_{rr} + \frac{2}{r} u_r + \frac{1}{r^2 \sin \rho} \left(\sin \rho \cdot \frac{\partial u}{\partial \rho} \right)_\rho + \left(\frac{1}{r^2 \sin^2 \rho} \cdot \frac{\partial u}{\partial \theta} \right)_\theta - u_{tt}$$

and the following differential identity holds

$$(10) \qquad v_r L(u) + u_r M(v)$$

$$= -(u_t v_r + u_r v_t)_t + (u_t v_t + u_r v_r)_r + \frac{v_r}{r^2 \sin \rho} \left(\sin \rho \cdot \frac{\partial u}{\partial \rho} \right)_\rho + v_r \left(\frac{1}{r^2 \sin^2 \rho} \cdot \frac{\partial u}{\partial \theta} \right)_\theta$$

where

$$(11) \qquad M(v) = v_{rr} - \frac{2}{r} v_r - v_{tt}$$

Let $C_{\overline{t}}$ denote the four dimensional conical volume with vertex $(\overline{x}, \overline{y}, \overline{z}, \overline{t})$ which is determined by the inequalities

$$(x - \overline{x})^2 + (y - \overline{y})^2 + (z - \overline{z})^2 \leqq (\overline{t} - t)^2, \qquad t \geqq 0$$

or equivalently, by

$$0 \leqq \rho \leqq \pi, \quad 0 \leqq \theta \leqq 2\pi, \quad 0 \leqq r \leqq \overline{t} - t, \quad 0 \leqq t \leqq \overline{t}$$

and denote by $B_{\overline{t}}$ the three dimensional base of $C_{\overline{t}}$, which is determined by

$$(x-\bar{x})^2 + (y-\bar{y})^2 + (z-\bar{z})^2 \leqq \bar{t}^2, \quad t = 0.$$

Choosing

$$v = v(r,t) = (\bar{t}-t)^2 - r^2 \tag{12}$$

which satisfies $M(v) = 0$, and integrating the differential identity (10) multiplied by $\frac{1}{r^2}$ (recall that the volume element $dx\,dy\,dz\,dt = r^2 \sin\rho\,dr\,d\rho\,d\theta\,dt$) over the volume $C_{\bar{t}}$, one obtains

$$\int_{\rho=0}^{\pi} \int_{\theta=0}^{2\pi} \int_{t=0}^{\bar{t}} \int_{r=0}^{r+t=\bar{t}} [v_r L(u) + u_r M(v)] \sin\rho \; dr\, d\rho\, d\theta\, dt$$

(13)

$$= \int_{\rho=0}^{\pi} \int_{\theta=0}^{2\pi} \left\{ \int_{r=0}^{\bar{t}} [u_t v_r + u_r v_t]_{t=0} dr \right\} \sin\rho \; d\rho\, d\theta$$

$$- \int_{\rho=0}^{\pi} \int_{\theta=0}^{2\pi} \left\{ \int_{t=0}^{\bar{t}} [u_t v_t + u_r v_r]_{r=0} dt \right\} \sin\rho \; d\rho\, d\theta$$

Now, $L(u) = 0$, $M(v) = 0$, and

$$v_r\big|_{r=0} = 0, \qquad v_r\big|_{t=0} = -2r$$

$$v_t\big|_{r=0} = -2(\bar{t}-t), \qquad v_t\big|_{t=0} = -2\bar{t}$$

therefore, returning to $x, y\; z, t$ variables, (13) yields

$$\int_{B_{\bar{t}}} [u_t(x,y,z,0) \cdot \frac{1}{r} + u_r(x,y,z,0) \cdot \frac{\bar{t}}{r^2}] \; dx\, dy\, dz$$

$$= 4\pi \int_{t=0}^{\bar{t}} u_t(\bar{x},\bar{y},\bar{z},t) \cdot (\bar{t}-t) \; dt$$

$$= 4\pi \int_{0}^{\bar{t}} dt_1 \left[\int_{0}^{t_1} u_t \; (\bar{x},\bar{y},\bar{z},t_2) \; dt_2 \right]$$

Differentiating with respect to $\bar{t}$ leads to the final formula

$$4\pi u(\bar{x},\bar{y},\bar{z},\bar{t}) = 4\pi u(\bar{x},\bar{y},\bar{z},0)$$

(14)
$$+ \frac{\partial}{\partial \bar{t}}\left[\int_{B_{\bar{t}}} u_t(x,y,z,0)\cdot\frac{1}{r}\cdot dx\,dy\,dz\right]$$

$$+ \frac{\partial}{\partial \bar{t}}\left[\int_{B_{\bar{t}}} u_r(x,y,z,0)\cdot\frac{\bar{t}}{r^2}\cdot dx\,dy\,dz\right]$$

This formula for the solution of the Cauchy problem (8) can be easily shown to coincide with the more familiar Poisson formula for the solution. The preceding argument has established the following uniqueness theorem: if the Cauchy problem (8) has a solution u which possesses continuous second derivatives on $t > 0$, and continuous first derivatives on $t \geq 0$), then u is given by equation (14).

§2. SINGULAR CAUCHY PROBLEM FOR THE EULER-POISSON-DARBOUX EQUATION

The Cauchy problem in question requires the determination of a function $u(x_1,\ldots,x_n,t)$ which satisfies the Euler-Poisson-Darboux equation

(15)
$$\Delta u = u_{tt} + \frac{k}{t}\,u_t$$

(where $\Delta v = \sum_{i=1}^{n} \frac{\partial^2 v}{\partial x_i^2}$ is the Laplacian, and k is a real number) and which meets the initial conditions

(16)
$$u(x_1,\ldots,x_n,0) = f(x_1,\ldots,x_n), \qquad u_t(x_1,\ldots,x_n,0) = 0$$

with f a given function. When $k = 0$ in (15), the equation reduces to the wave equation (7); while for $k \neq 0$, the coefficient k/t is infinite on the plane $t = 0$, the carrier of the Cauchy data.

For k any real number, this singular Cauchy problem was first solved by A. Weinstein [22], who employed what he termed the "method of recurrence" and a generalized method of descent. Let the solution of the Cauchy problem (15), (16) be denoted by u^k. Weinstein shows that for $k \neq -1, -3, \ldots$ the Cauchy problem for u^k can be reduced to a Cauchy problem for u^{k+2} by means of the following recurrence formulas:

$$(17)\qquad u^{k+2}(x,t) = \frac{u_t^k(x,t)}{t}; \quad u^{k+2}(x,0) = \frac{\Delta u^k(x,0)}{k+1}, \quad u_t^{k+2}(x,0) = 0$$

where x stands for $(x_1,\ldots,x_n)$. Moreover, for arbitrary solutions u^k of equation (15), there is the correspondence principle

$$(18)\qquad u^k = t^{1-k}u^{2-k}$$

which enables one to obtain from a solution u^k of (15) a solution of the same equation with parameter value $2 - k$. Formula (18) is used in discussing the uniqueness of the solution. Weinstein first solves explicitly the Cauchy problem (15), (16) for $k \geqq n-1$, and then, by a process which consists essentially in a repeated application of the relation (17), obtains a formula for the solution when $k \geqq n-1$.

It is well known that for $k = n-1$ the Cauchy problem (15), (16) admits a unique solution, which is given by a generalization of a formula of Poisson (see Asgeirsson [1]):

$$(19)\qquad u(x_i,t) = \frac{1}{\omega_n} \int_{\sum_{j=1}^{n} \beta_j^2=1} f(x_i + t\beta_i)\,d\omega_n$$

where $d\omega_n$ is the surface element of the unit sphere in n dimensional Euclidean space, and

$$\omega_n = 2\pi^{n/2}/\Gamma\left(\tfrac{n}{2}\right)$$

is the surface area of this sphere. Using this formula and Hadamard's "method of descent" [10], an explicit formula for the solution u^k of (15), (16):

$$(20)\qquad u^k(x_i,t)$$

$$= \frac{\omega_{k+1-n}}{\omega_{k+1}} \int_{\sum_{j=1}^{n} \beta_j^2 \leq 1} f(x_1+\beta_1 t,\ldots,x_n+\beta_n t)\,(1-\beta_1^2-\ldots-\beta_n^2)^{\frac{k-n-1}{2}}\,d\beta_1\ldots d\beta_n$$

may be obtained when k is any integer of the sequence $n, n+1, \ldots,$ and this formula for u^k is then readily verified to be the solution of (15), (16) for any real $k > n - 1$. This procedure may be called a generalized method of descent. By means of this explicit formula (20) and the recurrence relation (17), Weinstein obtains an explicit formula for the solution u^k of (15), (16) for all values of k with the exception of $-1, -3, -5, \ldots$. For the sake of definiteness, suppose that $n - 3 < k < n-1$. Then Weinstein's formula is, by (17), simply

$$u^k(x,t) = \int_0^t r u^{k+2}(x,r)\, dr + f(x)$$

where u^{k+2} is given by (20), using the initial values

$$\frac{\Delta u^k(x,0)}{k+1}$$

from (17). Miss Davis [5] gave another form to this solution which, in the general case $k < n-1$, $k \neq -1, -3, \ldots,$ $k + 2m > n-1$, is given by

$$u^k(x,t) = \frac{1}{(k+1)(k+3)\ldots(k+2m-1)} \sum_{j=0}^{m} B_{m,m-j}\, t^j \frac{\partial^j}{\partial t^j} L(x,t,f)$$

where

$$L(x,t;f)$$

$$= \frac{\omega_{k+2m+1-n}}{\omega_{k+2m+1}} \int_{\sum_{i=1}^{n} \alpha_i^2 \leq 1} f\left(x_i+\alpha_i t\right)\left(1-\alpha_1^2-\ldots-\alpha_m^2\right)^{\frac{k+2m-1-n}{2}} d\alpha_1 \ldots d\alpha_n$$

with explicitly known constants B.

J. B. Diaz and H. F. Weinberger [9], obtained the solution of the Cauchy problem (15), (16), using again as a starting point the same explicit formula (20) for the solution for $k > n-1$ which was employed by Weinstein. The definite integral in (20) diverges for $k < n-1$, and their method consists in finding the analytic continuation of this integral for $k < n-1$. The resulting formula for the solution u^k of (15), (16) is valid for $k \neq -1, -3, -5, \ldots$ and again yields as a special case a solution of the corresponding problem for the wave equation (7). To avoid misunderstanding, it should be noted that this method differs from that used by M. Riesz [17], [18] (see also F. Bureau [3]) for solving regular Cauchy problems, although both methods employ analytic continuation of definite integrals. M. Riesz uses a modified fundamental solution depending on a parameter α, and Green's identity, and obtains the solution by analytic continuation with respect to α. In the present method, the family of differential equations (15) depending on the parameter k is considered, and the solution of the singular Cauchy problem is obtained by analytic continuation in k. Neither a fundamental solution nor a Green's identity is used. Both methods involve the analytic continuation of integrals which are related to the Riemann-Liouville integral. However, it should be kept in mind that the analytic continuation of divergent integrals, based on similar principles but for different purposes, occurs already in Cauchy [4]. The final formula for $k < n-1$, $k \neq -1, -3, \ldots$, is the following:

$$u^k(x,t) = \frac{2\,\Gamma(\frac{k+1}{2})}{\Gamma(\frac{k+2-n}{2})\Gamma(\frac{n}{2})} \int_0^1 N_p(x_i,\alpha,t;f)\,(1-\alpha^2)^{\frac{k-n-1}{2}}\,\alpha^{n-1}\,d\alpha$$

$$+ \frac{\Gamma(\frac{k+1}{2})}{\Gamma(\frac{n}{2})} \cdot \sum_{j=0}^{p} \frac{(-t)^j}{j!} \frac{\partial^j M}{\partial t^j}(x_i,t;f) \sum_{l=0}^{j} (-1)^l \binom{j}{l} \frac{\Gamma(\frac{n+l}{2})}{\Gamma(\frac{k+l+1}{2})}$$

where

$$M(x_i,r;f) = \frac{1}{\omega_n} \int_{\sum_{j=1}^{n} \beta_j^2 = 1} f(x_i+r\beta_i)\,d\omega_n, \qquad r \geqq 0$$

and

$$N_p(x_i,\alpha,t;f) = M(x_i,\alpha t;f) - \sum_{j=0}^{p} \frac{t^j(\alpha-1)^j}{j!} \frac{\partial^j}{\partial t^j} M(x_i,t;f)$$

The odd negative integer values of the parameter k play an exceptional role in the solution of the Cauchy problem (15), (16). A solution of the Cauchy problem for these exceptional values of k was given by Diaz and Weinberger [9], after Weinstein [22] had pointed out the particular nature of polyharmonic initial values in this problem. The result is that, if $f(x_1, \ldots, x_n)$ has continuous derivatives of order not less than the maximum of the two numbers $\frac{1}{2}(n+3-k)$ and $3 - k$, there is a solution of (15), (16). The derivative of order $1 - k$ of the solution with respect to t becomes infinite like $\log t$ at $t = 0$ unless $f(x_1, \ldots, x_n)$ is polyharmonic of order $\frac{1}{2}(1-k)$, in which case the solution is just a polynomial in t. It should be noticed that for these exceptional values $k = -1, -3, \ldots,$ the solution does not depend continuously on the initial data, inasmuch as the slightest (no matter how smooth) deviation from the proper polyharmonic initial values produces a logarithmic infinity in t in certain derivatives of the solution. In the case $n = 1$, $k = -1, -3, \ldots,$ another solution of (15), (16) was given recently by Blum [2], under the weaker differentiability assumption that $f(x_1)$ has continuous derivatives of order $\frac{1}{2}(5-k)$. Blum also clarified the question of the uniqueness of the solution when $n = 1$ and k is any negative number, not necessarily an integer.

§3. GENERALIZED AXIALLY SYMMETRIC POTENTIAL THEORY

This theory deals with the elliptic equation

$$\frac{\partial^2 u}{\partial x_1^2} + \cdots + \frac{\partial^2 u}{\partial x_n^2} + \frac{\partial^2 u}{\partial y^2} + \frac{k}{y} \frac{\partial u}{\partial y} = 0 \tag{21}$$

which corresponds formally to the hyperbolic Euler-Poisson-Darboux equation (15) by means of the substitution $y = it$. For positive integral values of k, equation (21) is Laplace's equation in $n + k + 1$ dimensions when the solution u is axially symmetric in the last $k + 1$ variables, i.e.

$$u\left(x_1, \ldots, x_n, x_{n+1}, \ldots, x_{n+k+1}\right) = u\left(x_1, \ldots, x_n, \left[x_{n+1}^2 + \cdots + x_{n+k+1}^2\right]^{1/2}\right)$$

and

$$y = \left[x_{n+1}^2 + \cdots + x_{n+k+1}^2\right]^{1/2}$$

This equation may be considered as a simple example of an elliptic equation with rational coefficients which become infinite for real values of the independent variables. This type of differential equation is also of particular interest because it arises in a variety of seemingly disconnected fields in pure and applied mathematics (see e.g. A. Weinstein [20] and L. E. Payne [16]). Inasmuch as this theory was recently reviewed in a comprehensive manner by A. Weinstein [23] in an address delivered to the American Mathematical Society, only some developments since that address will be mentioned.

J. B. Diaz and A. Weinstein [8] have shown that a fundamental solution $u_b^{(k)}$ for the equation (21) with a singularity at the point $x_1 = x_2 = \dots = x_n = 0$, $y = b > 0$ is given by the formula

$$(22) \qquad u_b^{(k)} = \int_0^{\pi} \frac{\sin^{k-1}\alpha \, d\alpha}{\left(\sum_{i=1}^{n} x_i^2 + b^2 + y^2 - 2\,by \cos\alpha \right)^{\frac{k+n-1}{2}}}$$

for $k > 0$. A similar formula for a fundamental solution of (21), for $k \leqq 1$, is obtained by an application of the recurrence relation (18), (with y replacing t) which holds for solutions of (21). These formulas constitute an extension to (21) of formulas for a fundamental solution of the equation

$$\phi_{xx} + \phi_{yy} + \frac{p}{y}\phi_y = 0$$

with singularity at $x = 0$, $y = b > 0$, which were given by Weinstein [19]. The case $p = 1/3$ corresponds to Tricomi's equation. It is worth noting that the substitution $y = it$, $b = i\beta$, $\beta > 0$ in (22) yields immediately a fundamental solution for the Euler-Poisson-Darboux equation (15), and a similar remark applies to the corresponding formula for the fundamental solution for $k \leqq 1$. However, this fundamental solution for (15), despite its obvious importance, was not used in Section 2 in solving the singular Cauchy problem (15), (16) for the Euler-Poisson-Darboux equation.

In connection with the theory of equation (21), far reaching results have been recently obtained by A. Huber. For example, Huber has obtained the following theorem concerning the solutions of (21) for $k < 1$ (this theorem is an extension to (21) of the Phragmén-Lindelöf theorem for harmonic functions, as formulated by M. H. Heins [11]: Let H denote the half-space $y > 0$, S the hyperplane $y = 0$, and let u be a solution of (21) with $k < 1$, defined in H and satisfying

$$\limsup_{P \to Q} u(P) \leqq 0$$

for all points Q on S. Then it follows that:

(1) the limit

$$\lim_{r \longrightarrow \infty} \frac{m(r)}{r^{1-k}} = \alpha$$

(where $m(r) = \sup u(P)$ for P in the intersection of H and the sphere $\sum_{i=1}^{n} x_i^2 + y^2 = r^2$) always exists (either finite or infinite);

(2) $\alpha \geqq 0$; and

(3) throughout H the inequality

$$u(x_1, \ldots, x_n, y) \leqq \alpha y^{1-k} \tag{23}$$

holds.

Furthermore, if the equality sign holds in (14) at a single point of H, then $u \equiv \alpha y^{1-k}$.

Another result of Huber [12] concerning solutions of (21) runs as follows: Let G be a region lying in the half-space H, and suppose that the boundary of G contains an open subset of S. Then, if u is a solution of (21) taking the boundary value 0 on the portion of the boundary of G lying in S, one can assert that:

(1) for $k \geqq 1$, $u \equiv 0$ throughout G;

(2) for $k < 1$, $u = y^{1-k}v$, where v is analytic on $y = 0$ and satisfies (21) with k replaced by $2 - k$. The proof involves the use of the identification principle for solutions of (21) which are even in y and analytic on $y = 0$ (which was given originally by Weinstein [19] for equation (21) with $n = 1$), and a theorem on removable singularities on $y = 0$.

BIBLIOGRAPHY

[1] ASGEIRSSON, L., "Über eine Mittelwerteigenschaft von Lösungen homogener linearen partiellen Differentialgleichungen zweiter Ordnung mit konstanten Koeffizienten," Mathematische Annalen 113 (1937), 321-346.

[2] BLUM, E., "On the Euler-Poisson-Darboux equation for negative values of the parameter," (in press).

[3] BUREAU, F., "Intégrales de Fourier et problème de Cauchy," Annali di Matematica Pura ed Applicata, Series 4 32 (1951), 205-233.

[4] CAUCHY, A. L., Oeuvres complètes, Series 2 6, Gauthier-Villars, Paris, (1887), 78-88.

[5] DAVIS, RUTH M., Abstract, Meeting of the American Mathematical Society, April (1953), New York.

[6] DIAZ, J. B., and MARTIN, M. H., "Riemann's method and the problem of Cauchy. II. The wave equation in n dimensions," Proceedings of the American Mathematical Society 3 (1952), 476-483.

[7] DIAZ, J. B., and MARTIN, M. H., "Riemann's method for partial differential equations in two independent variables," Proceedings of the Eighth International Congress on Theoretical and Applied Mechanics, Istanbul, Turkey (1952).

BIBLIOGRAPHY

[8] DIAZ, J. B., and WEINSTEIN, A., "On the fundamental solutions of a singular Beltrami operator," (to appear in the R. von Mises Anniversary Volume).

[9] DIAZ, J. B., and WEINBERGER, H. F., "A solution of the singular initial value problem for the Euler-Darboux-Poisson equation," Proceedings of the American Mathematical Society, (in press).

[10] HADAMARD, J., Lectures on Cauchy's Problem in Linear Partial Differential Equations, Yale University Press, New York (1923).

[11] HEINS, M. H., "On the Phragmén-Lindelöf principle," Transactions of the American Mathematical Society 60 (1946), 238-244.

[12] HUBER, A., "On the uniqueness of generalized axially symmetric potentials," Annals of Mathematics (in press).

[13] LEWY, H., "Verallgemeinerung der Riemannschen Methode auf mehr Dimensionen," Nachrichten der Gesellschaft der Wissenschaften zu Göttingen. Mathematisch-Physikalische Klasse (1928), 118-123.

[14] MARTIN, M. H., The rectilinear motion of a gas," American Journal of Mathematics 45 (1943), 391-407.

[15] MARTIN, M. H., "Riemann's method and the problem of Cauchy," Bulletin of the American Mathematical Society 57 (1951), 238-249.

[16] PAYNE, L. E., "On axially symmetric flow and the method of generalized electrostatics," Quarterly of Applied Mathematics 10 (1952), 197-204.

[17] RIESZ, M., "L'intégrale de Riemann-Liouville et le problème de Cauchy pour l'équation des ondes," Conférence a la reunion internationale des mathématiciens tenue à Paris en Juillet 1937, Gauthier-Villars, Paris (1938), 153-170.

[18] RIESZ, M., "L'intégrale de Riemann-Liouville et le problème de Cauchy," Acta Mathematica 81 (1949), 1-223.

[19] WEINSTEIN, A., "Discontinuous integrals and generalized axially symmetric potential theory," Transactions of the American Mathematical Society 63 (1948), 342-354.

[20] WEINSTEIN, A., "The method of singularities in the physical and in the hodograph plane," Fourth Symposium in Applied Mathematics (in press).

[21] WEINSTEIN, A., "On the wave equation and the equation of Euler-Poisson," Fifth Symposium in Applied Mathematics (in press).

[22] WEINSTEIN, A., "Sur le problème de Cauchy pour l'équation de Poisson et l'équation des ondes," Comptes rendus de l'Académie des Sciences, Paris 234 (1952), 2584-2585.

[23] WEINSTEIN, A., "Generalized axially symmetric potential theory," Bulletin of the American Mathematical Society 59 (1953), 20-37.

XV. A BOUNDARY VALUE PROBLEM FOR THE WAVE EQUATION AND MEAN VALUE THEOREMS

M. H. Protter

§1. INTRODUCTION

For the wave equation in two independent variables

$$u_{xx} = u_{yy}$$

it is well known that in addition to the Cauchy problem a variety of boundary value problems can be solved. These are problems in which values of the unknown function are prescribed along two intersecting curves (one or both of which may be characteristics) and then the solution is determined in some "influence domain," see Goursat [4]. For hyperbolic equations in more than two variables the problems discussed are usually the Cauchy problem and the characteristic initial value problem. The latter is one in which values of the unknown function are prescribed on a characteristic surface. Soboleff [8] has discussed, for the wave equation in an arbitrary number of variables, the problem of prescribing boundary values along a conical time-like surface. The same problem has been treated recently by a different method by Gårding [3]. H. Lewy[1] has solved certain boundary value problems for general second order hyperbolic equations. It is the purpose of this paper to show how some boundary value problems other than those considered by Soboleff and Gårding may be solved for the wave equation in three variables. At the end of Section 2 a remark is made concerning the extension to the case of the wave equation in an arbitrary number of variables.

Let D be the domain bounded by the three surfaces

$$x^2 + y^2 = (z-z_0)^2, \qquad (x-x_0)^2 + (y-y_0)^2 = z^2, \qquad z = 0$$

where of course the fixed quantities x_0, y_0, z_0 must satisfy the condition $x_0^2 + y_0^2 < z_0^2$ if we are to have a domain. We denote by S_1 that part of the cone $(x-x_0)^2 + (y-y_0)^2 = z^2$ coinciding with D, by S_2 the part of $x^2 + y^2 = (z-z_0)^2$ coinciding with D and by S_3 the circle $x^2 + y^2 \leq z_0^2$

[1] Oral communication. I am indebted to H. Lewy for pointing out the feasibility of solving various boundary value problems for hyperbolic equations.

lying in the plane $z = 0$. Let $\phi_1(x,y)$, $\phi_2(x,y)$ be given continuously differentiable functions defined in the circle $x^2 + y^2 \leq z_0^2$. We seek a solution of the equation

$$u_{xx} + u_{yy} = u_{zz} \tag{1}$$

in D satisfying the boundary conditions

$$(2)\qquad \begin{aligned} u(x,y,0) &= \phi_1(x,y) \quad \text{for} \quad x,y \quad \text{on} \quad S_3 \\ u(x,y,z) &= \phi_2(x,y) \quad \text{for} \quad x,y,z \quad \text{on} \quad S_2 \end{aligned}$$

It is assumed that ϕ_1 and ϕ_2 coincide on the circle $x^2 + y^2 = z_0^2$, $z = 0$. We remark that it is not necessary to define ϕ_2 in the entire circle but only in the domain which is the projection of S_2 on the xy-plane. This solution will be found by an application of a generalization of Riemann's method due to Martin [7] and by use of a mean value theorem of Asgeirsson [1]. If instead of (2) data are prescribed along the characteristic surfaces S_1 and S_2 a solution may easily be obtained by solving the appropriate characteristic initial value problems. Also if boundary values are prescribed on S_1 and S_3 then a solution may be found by essentially the same method as that described below.

THEOREM 1. There exists a unique solution $u(x,y,z)$ of (1) in D satisfying the boundary conditions (2).

§2. PROOF OF THEOREM 1

We first consider the case $x_0 = y_0 = 0$ and without loss of generality we select $z_0 = 1$. To illustrate the method of proof we will first obtain u at a point on the z-axis. That is, we will determine $u(0,0,z)$, $0 < z \leq 1$. The first step will be to find the mean value of $u(x,y,z)$ taken over a circle on S_1 parallel to the plane $z = 0$. If we introduce characteristic coordinates α, β, ϕ by the relations

$$x = \tfrac{1}{2}(\alpha-\beta)\cos\phi, \qquad y = \tfrac{1}{2}(\alpha-\beta)\sin\phi, \qquad z = \tfrac{1}{2}(\alpha+\beta)$$

then (1) becomes

$$Lu \equiv u_{\alpha\beta} - \frac{1}{2(\alpha-\beta)}(u_\alpha - u_\beta) - \frac{u_{\phi\phi}}{(\alpha-\beta)^2} = 0 \tag{3}$$

Let P be a point on S_1 and C_P the circle on S_1 through P parallel to the xy-plane. If S_P denotes the characteristic (retrograde) cone containing C_P then a subdomain of D, say D_P is formed by parts of the four surfaces S_1, S_2, S_3 and S_P. We consider the operator "associated with Lu" (see [3], p. 247),

$$Mv \equiv v_{\alpha\beta} + \frac{1}{2(\alpha-\beta)}(v_\alpha - v_\beta) \tag{4}$$

In characteristic coordinates the domain D_P is transformed into a domain D_P^* and we examine the expression

$$\iiint_{D_P^*} [(v_\beta - v_\alpha)Lu - (u_\beta - u_\alpha)Mv]d\alpha\, d\beta\, d\phi \tag{5}$$

In a rectangular coordinate system with α, β, ϕ axes, the domain bounded by the cone $x^2 + y^2 = (z-1)^2$ and the unit circle in the xy-plane will be an infinite wedge. However, since we are interested in solutions of (3) which are periodic in ϕ with period 2π we may confine our attention to that part of the wedge lying between the planes $\phi = 0$ and $\phi = 2\pi$. The domain D_P^* is then a figure bounded by the six planes: $\phi = 0$, $\phi = 2\pi$, $\beta = 0$, $\alpha = -\beta$, $\alpha = 1$, $\alpha = K_P$; the quantity K_P is a positive constant less than one and depends on the height of the point P above the xy-plane. The integral (5) may be transformed into the surface integral

$$\iint_{B_P} \left[u_\beta v_\beta\, d\beta\, d\phi - u_\alpha v_\alpha\, d\alpha\, d\phi + u_\phi \frac{v_\alpha - v_\beta}{(\alpha-\beta)^2} d\alpha\, d\beta\right] \tag{6}$$

where B_P is the boundary of D_P^*. If u is a solution of (3) and v a solution of $Mv = 0$, (5) and therefore (6) will vanish. The integral (6) is to be evaluated over the six planar domains described above. First we note that the integrals taken over $\phi = 0$ and $\phi = 2\pi$ will cancel each other. This follows from the fact that v is independent of ϕ, u is periodic and the normals to these planes are selected with opposite signs. We therefore have

$$0 = \int_0^{2\pi}\int_1^{K_P} [-u_\alpha v_\alpha]_{\beta=0} d\alpha\, d\phi + \int_0^{2\pi}\int_{K_P}^{1} [u_\beta v_\beta - u_\alpha v_\alpha]_{\alpha=-\beta} d\alpha\, d\phi$$

$$-\int_0^{2\pi}\int_0^{-K_P} [u_\beta v_\beta]_{\alpha=K_P}\, d\beta\, d\phi - \int_0^{2\pi}\int_{-1}^{0} [u_\beta v_\beta]_{\alpha=1}\, d\beta\, d\phi = I_1 + I_2 + I_3 + I_4$$

Since v is independent of ϕ we determine it in the trapezoid (in the plane $\phi = 0$) bounded by the lines $\beta = 0$, $\alpha = -\beta$, $\alpha = K_P$, $\alpha = 1$. Then v is defined throughout D_P^*. In such a trapezoid we can solve the boundary value problem for $Mv = 0$ in which values are assigned along the lines $\beta = 0$, $\alpha = -\beta$ and $\alpha = K_P$. See, for example, [2], p. 155 ff. Also problems of this kind for equations similar to $Mv = 0$ have been considered by Hadamard [5], [6]. Thus the quantity v is at our disposal and the conditions we impose are

$$v_\alpha = 1 \text{ on } \beta = 0, \quad v_\beta = -1 \text{ on } \alpha = K_P, \quad v_\alpha = v_\beta \text{ on } \alpha = -\beta$$

The solution of $Mv = 0$ subject to these conditions may introduce a discontinuity in v_β along the line $\beta = -K_P$. However, v_β does not occur in the second term of (6) above, the only possible one affected by this discontinuity. We can perform one of the integrations in I_1 and obtain

$$-\frac{1}{2\pi} I_1 = \frac{1}{2\pi}\int_0^{2\pi} [u]_{\alpha=K_P,\beta=0}\,d\phi - \frac{1}{2\pi}\int_0^{2\pi} [u]_{\alpha=1,\beta=0}\,d\phi = I_{11} + I_{12}$$

The integral I_{12} when considered in xyz-space is merely the mean value of u taken over the circle common to the surfaces S_1 and S_2 while I_{11} is the mean value of u taken over the circle C_P and is the quantity we wish to determine. For I_2 we have

$$-\frac{1}{2\pi} I_2 = \frac{1}{2\pi}\int_0^{2\pi}\int_{K_P}^{1} v_\alpha (u_\alpha - u_\beta)\,d\alpha\,d\phi$$

However, along the line $\alpha = -\beta$, $\phi = \text{const.}$ we find $du = (u_\alpha - u_\beta)d\alpha$ and since by hypothesis u is known on the plane $\alpha = -\beta$ the integral I_2 is completely known. The integral I_3 becomes

$$-\frac{1}{2\pi} I_3 = \frac{1}{2\pi}\int_0^{2\pi} [u]_{\alpha=K_P,\beta=0}\,d\phi - \frac{1}{2\pi}\int_0^{2\pi} [u]_{\alpha=K_P,\beta=-K_P}\,d\phi$$

The first term on the right is identical with I_{11} while the second term is a known quantity in terms of the given data. (In xyz-space it is the mean value of u over a circle in the xy-plane with center at the origin.) The remaining integral I_4 is known since u is given in the plane $\alpha = 1$ (i.e. on the surface S_2 in xyz-space); we note that u_β is known in the plane $\alpha = 1$ if u is. Since $I_1 + I_2 + I_3 + I_4 = 0$ we get an expression for I_{11} in terms of the given data.

Once the value of I_{11} is known, the value of u on the z-axis can be found. The retrograde cone through C_P has its vertex on the z-axis and it is at this point (0,0,z) that u can be found. To do this we apply Asgeirsson's mean value theorem ([1], p. 421) which for any solution w of equation (1) states that

$$\int_{\Omega_2} w(x + \beta_1 t,\ y + \beta_2 t,\ 0)d\omega_2 = \int_{\Omega_2} w(x,\ y,\ \beta_1 t)d\omega_2$$

where β_1, β_2 are unit vectors, Ω_2 is the unit circle $\beta_1^2 + \beta_2^2 = 1$ and

$$d\omega_2 = \frac{d\beta_1}{\sqrt{1 - \beta_1^2}}$$

is an element of arc length on Ω_2. If we employ this theorem for a point (0,0,z) on the z-axis we get in our case

$$(7) \qquad \frac{1}{2\pi} \int_0^z \frac{u(0,\ 0,\ \tau)d\tau}{(z^2 - \tau^2)^{1/2}} = \frac{1}{2\pi} \int_{C_P} u(x,\ y,\ \tfrac{z}{2})dS$$

However the right side is merely the integral I_{11} which is known in terms of the given data. Thus (7) is an Abel integral equation and as such can be solved for $u(0, 0, z)$, $0 < z \leq 1$. Thus the solution is known on the z-axis.

The next step is the determination of the solution at all points in the domain between the characteristic cones $x^2 + y^2 = (z-1)^2$ and $x^2 + y^2 = z^2$, that is, those points lying above S_1 and below $x^2 + y^2 = (z-1)^2$. Let P_1 be such a point. It is no loss of generality to set $y_1 = 0$ as this merely involves a rotation. Thus, suppose P_1 has coordinates $(x_1,0,z_1)$, $x_1 < z_1 \leq 1$. We introduce new characteristic coordinates

$$x - x_1 = \tfrac{1}{2}(\alpha-\beta)\cos\phi, \qquad y = \tfrac{1}{2}(\alpha-\beta)\sin\phi, \qquad z = \tfrac{1}{2}(\alpha+\beta)$$

The retrograde cone with vertex at P_1 will intersect S_1 in an ellipse; we denote by P the point on this ellipse with the smallest z-value, i.e. the point with coordinates

$$\left(\frac{x_1 - z_1}{2},\ 0,\ \frac{z_1 - x_1}{2}\right)$$

Let C_P be the circle on this retrograde cone parallel to the xy-plane and passing through P. This circle will lie entirely in D (except for the point P). The portion of the surface of this retrograde cone lying between C_P and the xy-plane we call S_P. We now construct the characteristic (direct) cone H through C_P having its vertex below the xy-plane. Then H intersects S_2 in an ellipse and we denote the portion of H between C_P and this ellipse by S_Q. The surfaces S_P, S_Q and the appropriate portions of S_2 and S_3 cut off by S_P and S_Q which we denote by S_2' and S_3' respectively enclose a sub-domain of D which we again denote by D_P; its transform in the space of characteristic coordinates we call D_P^*. We again form the expression (5) for the domain D_P^* and repeat the analysis as before. In α, β, ϕ space the domain is bounded by $\phi = 0$, $\phi = 2\pi$, $\alpha = \text{const.}$, $\beta = -x_1$, $\alpha = -\beta$, and (in place of the plane $\alpha = 1$) a characteristic surface corresponding to S_2' which we denote by T_2'. Thus the integrals I_1, I_2, I_3 are exactly the same while the integral I_4 is again taken over a characteristic surface on which u is known by hypothesis. Since T_2' is characteristic for equation (3) and we know u on T_2' we can find u_β on this surface. From the fact that on any characteristic surface u_β satisfies a first order partial differential equation it follows that u_β will be completely determined subject to proper initial conditions. Since u is prescribed on S_2 and S_3 the first derivatives of u are known in two independent directions on the circle common to S_2 and S_3. Thus u_β is known on this circle and these values may be used as initial conditions. Hence we again determine I_{11} in terms of known quantities and solve the corresponding Abel equation (7) to obtain u at the point P_1. In particular, the solution is now known over the entire surface of the cone $x^2 + y^2 = (z-1)^2$, $0 \leq z \leq 1$, and by a well known theorem this determines u in the interior. If the point x_0, y_0 is located elsewhere in the unit circle it is clear that the arguments above can be applied in a completely analogous way. From the fact that the solution is determined explicitly we see that if ϕ_1 and ϕ_2 vanish the function u obtained will vanish identically. It can be verified in the same way as for the case of the characteristic initial value problem that the function u obtained actually satisfies the wave equation (1). (See [1]).

For the wave equation in more than three variables similar problems may be solved. In this case use would have to be made of the extension of Martin's result given by Diaz and Martin [2].

§3. MEAN VALUE THEOREMS

The result of the previous section allows us to establish a mean value theorem which may be considered as a generalization of the theorem establishing the solution of the Abel integral equation. For functions which are continuous for $x \geq 0$ and differentiable at the origin the solution of an Abel equation may be interpreted as a mean value theorem. Suppose $f(x)$, $x \geq 0$, is imagined as a function of (x,y) which is independent of y. Suppose its mean value is known over all circles of the form $(x-a)^2 + y^2 = a^2$, $a \geq 0$, i.e., over all circles with center on the positive x-axis which are tangent to the y-axis. Then the determination of the function $f(x)$ is equivalent to solving the Abel integral equation

$$\int_0^a \frac{f(x)dx}{\sqrt{x}\,\sqrt{a - x}} = g(a)$$

where $g(a)$ is a known function.

This result may be generalized in the following way.

THEOREM 2. Let $F(x,y)$ be continuously differentiable in the unit circle K and suppose $P_0(x_0,y_0)$ is a fixed point in K. If the mean value of F is given as a continuously differentiable function $g(x,y)$ over all circles in K which are tangent to the boundary of K and do not contain the point x_0,y_0 then $F(x,y)$ is uniquely determined.

PROOF. The solution of the Cauchy problem for the wave equation is given by the formula

$$u(x,y,z) = \frac{1}{2\pi} \int_0^{2\pi}\int_0^z \frac{g_0(x + \rho \cos\phi,\ y + \sin\phi)}{(z^2 - \rho^2)^{1/2}} \rho\, d\rho\, d\phi$$

(8)

$$+ \frac{\partial}{\partial z} \frac{1}{2\pi} \int_0^{2\pi}\int_0^z \frac{f_0(x + \rho\cos\phi,\ y + \rho\sin\phi)}{(z^2 - \rho^2)^{1/2}} \rho\, d\rho\, d\phi$$

where $f_0 = u(x,y,0)$ and $g_0 = u_z(x,y,0)$ are Cauchy data defined in the unit circle $K : x^2 + y^2 \leq 1$, $z = 0$. Suppose first that $x_0 = y_0 = 0$ and consider the problem of the previous section. Then from the fact that f_0 and

$$u\left(x,y,1 - \sqrt{x^2 + y^2}\right), \qquad \left(\tfrac{1}{4} \leq x^2 + y^2 \leq 1\right)$$

are given, g_0 is already determined by the result of the previous section.

The above formula may be written in the form

$$\int_0^z \frac{h_0(x,\ y,\ \rho)d\rho}{(z^2 - \rho^2)^{1/2}}$$

(9)

$$= u(x,y,z) - \frac{\partial}{\partial z}\frac{1}{2\pi}\int_0^{2\pi}\int_0^z \frac{f_0(x + \rho\cos\phi,\ y + \rho\sin\phi)\rho\, d\rho\, d\phi}{(z^2 - \rho^2)^{1/2}}$$

where

$$h_0(x,y,\rho) = \int_0^{2\pi} g_0(x + \rho\cos\phi,\ y + \rho\sin\phi)d\phi$$

Thus $h_0(x,y,\rho)$ is the mean value of g_0 on a circle of radius ρ, center at (x,y). Consider the integral equation (9) for values of (x,y,z) on the cone $x^2 + y^2 = (z-1)^2$ for $\frac{1}{4} \leq x^2 + y^2 \leq 1$. Then by hypothesis the right side is known and (9) may be solved for $h_0(x,y,\rho)$. But then Theorem 1 states that g_0 is known. Hence, we conclude that given $h_0(x,y,\rho)$ we can obtain g_0.

If in Section 2 we solved the problem in which we prescribe boundary values on S_1 and S_3 we would be led to the following mean value theorem.

THEOREM 3. Let K, P_0 and $F(x,y)$ be as in Theorem 2. If the mean value of F is given as a continuously differentiable function $g(x,y)$ over all circles in K passing through P_0 then $F(x,y)$ is uniquely determined.

BIBLIOGRAPHY

[1] COURANT, R., and HILBERT, D., Methoden der Mathematischen Physik, Vol 2, Interscience Publishers, Inc., New York (1943).

[2] DIAZ, J. B., and MARTIN, M. H., "Riemann's method and the problem of Cauchy," II, Proceedings of the American Mathematical Society 3 (1952), 476-483.

[3] GÅRDING, L., "Le problème de Goursat pour l'équation des ondes," Comptes Rendus du Onzième Congrès de Mathématiciens Scandinaves, Tronheim (1949).

[4] GOURSAT, E., Cours d'Analyse Mathématique, III, 4th ed., Gauthier-Villars, Paris, (1927).

BIBLIOGRAPHY

[5] HADAMARD, J., "Sur un problème mixte aux dérivées partielles," Bulletin de la Société Mathématique de France, 31 (1903).

[6] HADAMARD, J., "Résolution d'un problème aux limites pour les équations linéaires du type hyperbolique, Bulletin de la Société Mathématique de France, 32 (1904).

[7] MARTIN, M. H., "Riemann's method and the problem of Cauchy," Bulletin of the American Mathematical Society, 57 (1951), 238-249.

[8] SOBOLEFF, S., Quelques problèmes nouveaux pour les équations aux dérivées partielles du type hyperbolique, Matematiceskii Sbornik, 11 (1942), 155-203.

Ingram Content Group UK Ltd.
Milton Keynes UK
UKHW032005260523
422428UK00001B/7